Springer Protocols Handbooks

For further volumes:
http://www.springer.com/series/8623

Springer Protocols Handbooks collects a diverse range of step-by-step laboratory methods and protocols from across the life and biomedical sciences. Each protocol is provided in the Springer Protocol format: readily-reproducible in a step-by-step fashion. Each protocol opens with an introductory overview, a list of the materials and reagents needed to complete the experiment, and is followed by a detailed procedure supported by a helpful notes section offering tips and tricks of the trade as well as troubleshooting advice. With a focus on large comprehensive protocol collections and an international authorship, Springer Protocols Handbooks are a valuable addition to the laboratory.

Microbial Biofilm Research

Comprehensive Protocols and Best Practices

Edited by

Abirami K

Department of Microbiology, SRM Arts and Science College, Kattankulathur, Tamil Nadu, India

Humana Press

Editor
Abirami K
Department of Microbiology
SRM Arts and Science College
Kattankulathur, Tamil Nadu, India

ISSN 1949-2448 ISSN 1949-2456 (electronic)
Springer Protocols Handbooks
ISBN 978-1-0716-5534-4 ISBN 978-1-0716-5535-1 (eBook)
https://doi.org/10.1007/978-1-0716-5535-1

This Humana imprint is published by the registered company Springer Science+Business Media, LLC, part of Springer Nature.
The registered company address is: 1 New York Plaza, New York, NY 10004, U.S.A.

Foreword

It gives me immense pleasure to write the foreword for this timely and insightful volume on *Microbial Biofilm Research.* Over the past few decades, our understanding of microorganisms has undergone a remarkable transformation. Once viewed primarily as free-floating, independent cells, microbes are now recognized as highly organized communities capable of sophisticated communication, adaptation, and survival. Biofilms represent one of the most fascinating expressions of this microbial intelligence.

Microbial biofilms are not merely surface-attached clusters of cells, they are dynamic, structured ecosystems embedded in self-produced extracellular matrices. These communities exhibit enhanced resistance to environmental stress, antimicrobial agents, and host immune responses. From clinical infections and chronic inflammatory conditions to environmental sustainability, water systems, aquaculture, and industrial processes, biofilms play both beneficial and detrimental roles. Their dual nature makes them a subject of immense scientific importance.

As someone working in the fields of toxicology, pharmacology, environmental sciences, molecular immunology, and molecular biology, I have witnessed how biofilm research intersects with multiple disciplines. Understanding biofilm-associated oxidative stress, host–pathogen interactions, antimicrobial resistance mechanisms, and ecological adaptations is crucial for developing innovative therapeutic strategies and sustainable environmental solutions. This book successfully bridges fundamental microbiology with applied research, offering readers a comprehensive perspective on these interconnected themes.

The contributors to this volume have presented current knowledge with clarity and scientific depth. The chapters thoughtfully address molecular mechanisms of biofilm formation, quorum sensing, extracellular polymeric substances, immune evasion strategies, and modern analytical approaches. Importantly, the strength of this book lies not only in its technical detail but also in its practical orientation guiding readers through experimental design, biofilm cultivation, quantification, imaging, molecular characterization, and data interpretation.

I am confident that this work will serve as a valuable resource for students, academicians, researchers, and professionals in microbiology, biotechnology, environmental sciences, medical sciences, and related fields. It not only consolidates foundational concepts but also inspires critical thinking and innovation.

I commend the editors and authors for their dedication and scholarly effort in bringing out this significant contribution to the scientific community. May this book stimulate further research and foster collaborative efforts in advancing our understanding of microbial biofilms and their far-reaching implications.

A. Jesu Arockia Raj, Ph.D.

Professor in Biotechnology
Faculty of Science and Humanities
SRM Institute of Science and Technology
Kattankulathur, Tamil Nadu, India
27 February 2026

Preface

Microbial biofilms represent one of the most widespread and resilient forms of life on Earth. From environmental niches and industrial systems to chronic infections in clinical settings, biofilms shape ecological balance, influence material integrity, and complicate therapeutic strategies. Over the past decades, advances in microscopy, molecular biology, and omics technologies have transformed our understanding of biofilm structure, function, and regulation. Yet, despite remarkable progress, experimental variability and methodological gaps continue to challenge reproducibility and cross-disciplinary collaboration.

This volume in the Springer Protocols series is designed to address that need.

The primary aim of this book is to provide carefully curated, step-by-step methodologies that reflect both foundational practices and emerging innovations in microbial biofilm research. The protocols included here have been selected to support researchers at different stages of expertise—from graduate students establishing their first biofilm assays to experienced investigators refining advanced analytical techniques. Each chapter emphasizes clarity, reproducibility, and practical insight, combining detailed experimental procedures with troubleshooting guidance and critical notes derived from hands-on laboratory experience.

Biofilm research is inherently interdisciplinary. It intersects microbiology, biomedical sciences, environmental engineering, materials science, and systems biology. Accordingly, this volume encompasses a wide range of methodologies: in vitro and in vivo biofilm models, imaging and quantitative analysis techniques, molecular and genetic approaches, high-throughput screening strategies, and methods for studying biofilm-host and biofilm-material interactions. Where appropriate, protocols highlight considerations for different microbial systems, including bacterial and fungal biofilms, recognizing the diversity of organisms and conditions under which biofilms develop.

A defining feature of this collection is its focus on methodological rigor. Biofilm experiments are particularly sensitive to variables such as surface properties, nutrient composition, hydrodynamics, and incubation parameters. Even subtle deviations can produce significant differences in architecture and phenotypic behavior. The contributors to this volume have therefore emphasized experimental design, controls, data interpretation, and limitations—elements that are essential for generating robust and reproducible findings.

We are deeply grateful to the contributing authors, whose expertise and commitment have made this book possible. Their willingness to share refined laboratory practices and practical insights reflects the collaborative spirit that drives progress in this field. We also acknowledge the editorial team at Springer for their guidance and support throughout the development of this volume.

It is our hope that this book will serve not only as a technical manual but also as a bridge between disciplines and generations of scientists. By standardizing and disseminating reliable protocols, we aim to facilitate reproducible research, foster innovation, and ultimately contribute to a deeper understanding of microbial biofilms and their impact on health, industry, and the environment.

We trust that readers will find this volume to be a valuable and enduring resource in their scientific endeavors.

Kattankulathur, Tamil Nadu, India ***Abirami K***

Acknowledgment

The preparation of this volume has been a collective effort, and we would like to express our sincere gratitude to all those who contributed to its completion.

First and foremost, we thank the chapter authors for generously sharing their expertise, time, and carefully optimized protocols. Their dedication to methodological rigor and clarity has shaped this book into a practical and reliable resource for the biofilm research community. We especially appreciate their patience in revising and refining their contributions to meet the standards of reproducibility and transparency that this series represents.

We are grateful to the editorial and production teams at Springer for their continuous guidance, professionalism, and support throughout the development of this volume. Their experience and attention to detail ensured a smooth review and publication process.

We also acknowledge the many students, postdoctoral researchers, and laboratory staff whose daily efforts in optimizing experiments, troubleshooting assays, and maintaining research facilities form the backbone of microbial biofilm research. Although not all are named individually, their contributions are reflected in the robustness of the protocols presented here.

Finally, we extend our appreciation to our respective institutions and funding agencies for providing the infrastructure and financial support necessary to advance research in microbial biofilms. The collaborative environment fostered within the scientific community continues to drive innovation and progress in this dynamic field.

We hope that this volume serves as a testament to the collective commitment of researchers dedicated to understanding and addressing the complexities of microbial biofilms.

Abirami K

Contents

Part IV Molecular and Genetic Methods

Part V Physical and Mechanical Methods

Part VI Culture-Based Methods

Part VII Spectroscopic and Chemical Analysis

Contributors

ABIRAMI J • *Department of Biochemistry, SriMad Andavan Arts and science College (Autonomous), Trichy, Tamil Nadu, India*

ABIRAMI K • *Department of Microbiology, SRM Arts and Science College, Kattankulathur, Chennai, Tamil Nadu, India*

AMRITHA SATHYA P • *Microbial Biotechnology Laboratory, Department of Microbiology, Rathinam College of Arts and Science, Coimbatore, Tamil Nadu, India*

MONIKA Y • *Microbial Biotechnology Laboratory, Department of Microbiology, Rathinam College of Arts and Science, Coimbatore, Tamil Nadu, India*

RAJESH PALANISAMY • *Department of Immunology and Molecular Microbiology, Texas Tech University Health Sciences Center, Lubbock, TX, USA*

PRABHA NAGARAM • *Department of Microbiology, SRM Arts and Science College, Kattankulathur, Chennai, Tamil Nadu, India*

PRIYA P K • *Department of Microbiology, SRM Arts and Science College, Kattankulathur, Chennai, Tamil Nadu, India*

RAJALAKSHMI M • *Microbial Biotechnology Laboratory, Department of Microbiology, Rathinam College of Arts and Science, Coimbatore, Tamil Nadu, India*

SRINIVASAN P • *Microbial Biotechnology Laboratory, Department of Microbiology, Rathinam College of Arts and Science, Coimbatore, Tamil Nadu, India*

SUJITH P • *Department of Microbiology, SRM Arts and Science College, Kattankulathur, Chennai, Tamil Nadu, India*

VENKATESH KUMARESAN • *Department of Molecular Microbiology and Immunology, The University of Texas at San Antonio, San Antonio, TX, USA*

VINUSHIYA S • *Department of Microbiology, SRM Arts and Science College, Kattankulathur, Chennai, Tamil Nadu, India*

Part I

Background and Significance

Chapter 1

Biofilm Formation and Its Significance in Biological Systems

Abirami K

Abstract

Microbial biofilms are structured communities of microorganisms embedded within a self-produced extracellular polymeric substance (EPS) matrix that adhere to various surfaces. The ability of biofilms to persist in diverse environments from medical devices and industrial pipelines to natural ecosystems has profound implications for human health, industry, and environmental processes. Accurate identification and characterization of biofilms are critical for understanding their formation, behavior, and resistance mechanisms. This book, *Microbial Biofilm Research*, provides a comprehensive exploration of state-of-the-art methods used for biofilm detection, imaging, and analysis. It highlights a range of advanced techniques, including microscopic methods (transmission and scanning electron microscopy), molecular approaches (fluorescent in situ hybridization, PCR-based assays), biochemical assays (crystal violet staining, EPS quantification), and emerging omics technologies (metagenomics, transcriptomics, proteomics). Furthermore, special attention is given to the challenges of detecting biofilms in complex and clinical environments, as well as strategies to improve sensitivity and specificity. By providing a detailed overview of biofilm identification methods, this work serves as an essential resource for researchers and practitioners aiming to advance biofilm science and develop effective control and treatment strategies.

Key words Extracellular polysaccharide, Microbial biofilm, PCR-based assay, Imaging analysis, Biofilm identification methods

1 Introduction

1.1 Definition of Biofilm

A biofilm is a highly structured and dynamic community of microorganisms that adhere to the both biotic (living) and abiotic (non-living) surfaces. The microbes attach to the surface and become embedded in self-produced matrix composed primarily of exopolysaccharide and extra polymeric substances (EPS). The matrix plays a crucial in stability and function of the biofilm. The biofilm matrix is made up of various biological molecules such as polysaccharides, proteins, nucleic acids, extracellular DNA, and lipids. It protects microbial cells from environmental threads such as antibiotics, disinfectants, and host immune response. The

Abirami K (ed.), *Microbial Biofilm Research: Comprehensive Protocols and Best Practices*, Springer Protocols Handbooks, https://doi.org/10.1007/978-1-0716-5535-1_1,

biofilm lifestyle allows microbes to survive in hostile environments, coordinate behavior, and resist antimicrobial agents significantly better than planktonic cell [1, 2]. Biofilms are not a random accumulation of cells, but highly organized structures that exhibit complex behaviors such as cell-to-cell communication, resource sharing, and division of labor. The EPS matrix, composed many materials, serves as a scaffold that binds the cell together and anchors the biofilm to the surface [2, 3].

1.2 Formation and Development of Biofilm on Surfaces

The development of a biofilm typically progresses through five distinct phases: (1) initial adhesion, (2) firm anchoring, (3) early development phase, (4) advanced development phase, and (5) cell dispersal. The regulation of this process depends on factors including nutrient availability, surface features, and intercellular signaling network like quorum sensing (Table 1) [4]. Biofilms are omnipresent, occurring in virtually all natural, clinical, and industrial environments. In natural ecosystems, they colonize surfaces such as rock in rivers, sediments in aquatic habitats, and plant tissues, where they contribute to the nutrient cycle, organic matter decomposition, and stabilization of sediments [5]. In these settings, biofilms serve as a microhabitat that harbors diverse microbial communities, including bacteria, algae, archaea, and protozoa, forming the basis of many aquatic food webs (*see* **Note 1**) (Fig. 1).

In clinical contexts, biofilm readily develops on human tissues, including dental surfaces, mucosal linings such as in chronic sinusitis or cystic fibrosis, lung infections, and chronic wounds. They colonize inert medical devices, such as urinary catheters, prosthetic joints, heart valves, and contact lens [6]. Biofilm-associated infections are notoriously difficult to treat because biofilm cells can exhibit up to thousand-fold greater resistance to antibiotics compared to their planktonic counterparts. Furthermore, the EPS matrix shields cells from immune defense, after leading to persistent, relapsing infections.

Table 1
Various stages of biofilm formation

Stage	Description & drivers
Initial attachment	Mediated by physical forces and bacterial appendages; reversible → irreversible
Microcolony & maturation	Growth supported by EPS production; architecture formation; regulated by signals
Dispersion	Triggered either actively (enzymes, stress) or passively (flow, shear)
Environmental & microbial factors	Include physicochemical conditions, signaling molecules (like c-di-GMP), quorum sensing, substrate mechanics, and biosurfactants

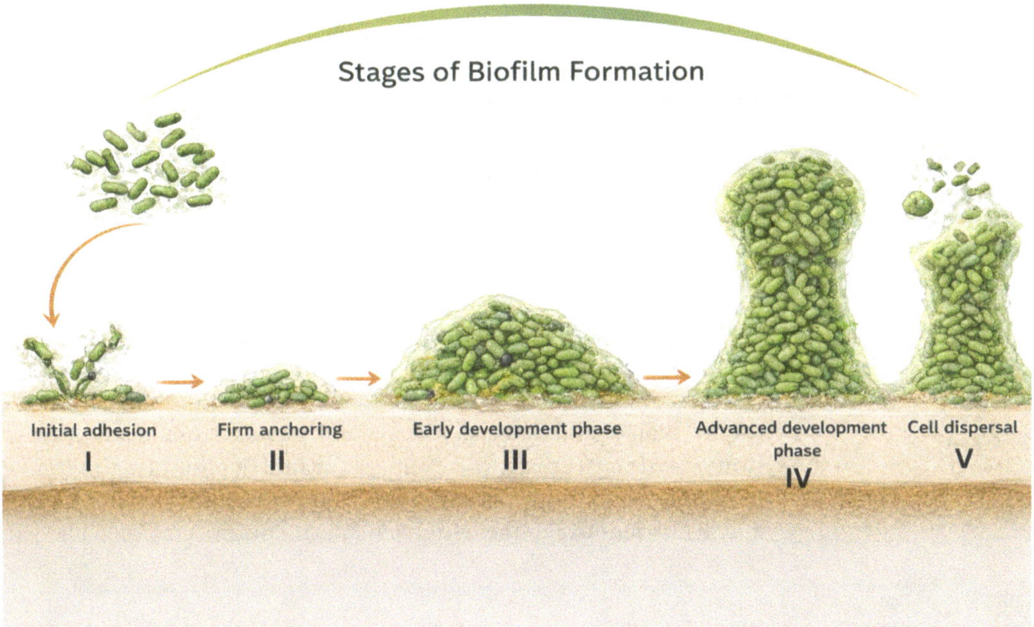

Fig. 1 Stages of biofilm formation. Figure shows the sequential stages involved in biofilm formation, a dynamic and highly regulated process that allows microorganisms to survive under adverse environmental conditions

In industrial and engineering systems, biofilms form on pipelines, cooling towers, ship hulls, and water distribution systems, where they can cause biofouling, microbially induced corrosion, and reduced operation efficiency. Biofilms in water treatment plants can be both beneficial and detrimental: beneficial in biofilters for removing contaminants, but problem when uncontrolled growth clogs the system or harbors pathogens (*see* **Note 2**) [7].

The persistent ability and adaptability of biofilms arise from their structural complexity, metabolic diversity, and well-structured communication by quorum sensing. The dual condition being both ecological beneficial and clinically problematic underscores the importance of understanding biofilm biology for application in medicine, environmental management, and industry [8].

Biofilm formation and dispersal are shaped by environmental factors such as temperature, pH, oxygen, nutrient levels, iron composition, and hydrodynamic and surface properties. These signals are processed through second messengers such as c-di-GMP and cAMP, which regulate adhesion, EPS production, motility, and dispersion. The quorum sensing system in bacteria coordinates community behaviors and is influenced by oxygen availability. Substrate mechanics and geometry affect biofilm structure and cell organization, while biosurfactants such as rhamnolipids maintain fluid channels and promote detachment [9, 10].

1.3 Microbial Diversity in Biofilm

Biofilms are diverse multi-kingdom communities comprising bacteria, archaea, fungi, protozoa, and algae that adapt to varied environments. In natural environments such as hot springs, thermophilic archaea dominate, while cyanobacteria and algae lead photosynthetic biofilms in cooler habitats. In nutrient-poor aquatic settings, biofilms play key roles in carbon, nitrogen, and methane cycle (Table 2). Industrial and wastewater systems often include archaeal biofilms for treatment processes. In medical and dental contexts, polymicrobial biofilm, such as those in dental plaque, harbor pathogenic bacteria, fungi, protozoa, and bacteriophages, with protozoa shaping bacterial diversity through predation [11–13].

1.4 Ecological and Functional Role

Biofilms support key ecosystem processes such as carbon fixation, nitrogen cycling, and methane oxidation by creating microenvironments that enable specialized metabolism and symbiotic interactions. In desert crusts, for example, cyanobacteria provide carbon to fungi, while fungi release minerals from rocks, forming a mutualistic nutrient exchange in nutrient-poor conditions [14]. The EPS

Table 2
Different microbes along with their biofilm-producing ability, functional role, and characterization of the biofilm

Microbial group	Example species	Typical biofilm environment	Notable role/feature
Bacteria	*Pseudomonas aeruginosa*	Medical devices, chronic wounds, water systems	Opportunistic pathogen, strong EPS producer, high antibiotic resistance
Bacteria	*Porphyromonas gingivalis*	Dental plaque, periodontal pockets	Keystone pathogen in periodontitis
Archaea	*Candidatus Caldiarchaeum subterraneum*	Geothermal hot springs	Thermophilic archaeon adapted to extreme heat
Archaea	*Methanosaeta* spp.	Anaerobic waste water treatment biofilms	Methanogenesis in anaerobic digestion
Fungi	*Candida albicans*	Oral cavity, medical catheters	Forms mixed biofilms with bacteria, antifungal resistance
Protozoa	*Acanthamoeba castellanii*	Water distribution systems, contact lenses	Predates bacteria, can harbor pathogens
Algae	*Oscillatoria* spp. (cyanobacteria)	Freshwater and marine biofilms	Photosynthetic primary producer
Viruses (Bacteriophages)	Various phages infecting *P. aeruginosa*	Within bacterial biofilms	Influence bacterial population dynamics, horizontal gene transfer

matrix in biofilms protects microbes from environmental stresses and greatly reduces antibiotic susceptibility. Biofilms also enhance stability and adaptability by promoting gene expression changes, horizontal gene transfer, and mutations, enabling long-term survival and diversification. Furthermore, biofilms enable microorganisms to survive extreme conditions such as high salinity, pressure, temperature extremes, and nutrient scarcity, by stabilizing cell attachment, trapping harmful ions, conserving resources, and offering physical protection, making them more resilient than free-living cells [15].

1.5 Medical Significance

Biofilms play a critical role in numerous infectious diseases, particularly chronic wounds, dental plaque, and infectious associated with medical devices such as catheters, prosthetic joints, and implants. Persistent biofilm-associated infections account for a larger proportion of chronic clinical problems due to the biofilm's protective matrix, which significantly limits antibiotics penetration, facilitates accumulation of degrading enzymes, and shelters slow-growing or dormant persister cells, contributing to resistance and recurrence [16]. Diagnosing these infections remains dependent on advanced imaging methods, which are not always feasible in routine care. Treatment difficulties are compounded by high antimicrobial tolerance, prompting need for novel therapeutic strategies. Emerging approaches include surface anti-adhesion technologies, innovative drug delivery systems, electric disruption methods, antimicrobial peptides, bacteriophages, and anti-biofilm agents such as metallic nanoparticles. All these approaches aim at preventing biofilm establishment and improving treatment efficacy against biofilms (*see* **Note 3**) [17].

2 Significance in Research and Innovation

Biofilms are at the forefront of microbiology, biotechnology, and public health research due to their central role in persistent infection, environmental processes, and industrial bioprocessing. Studies reveal that an estimated 65–80% of chronic human infections involve biofilm formation, underscoring their medical importance [18]. This has prompted innovative anti-biofilm strategies, including nano-delivery system such as liposomes, dendrimers, and metal organic frameworks that enhance drug penetration into biofilms and counter antibiotic resistance [19]. Alternative approaches also feature biosurfactants, quorum-sensing inhibitors, antimicrobial peptides, bacteriophages, and enzymes such as Dispersin B and DNase that can disrupt biofilm matrices and promote dispersion. Cutting-edge detection and analysis technologies are transforming our ability to study biofilms; specifically microfluidic-integrated biosensors, nanopore platforms, and lensless ptychographic image facilitate high-resolution, real-time monitoring, while photoacoustic flow cytometry offers rapid, sensitive detection of bacteria in blood and other complex samples [20–23].

Recent research is aggressively exploring next-generation antibiofilm strategies to overcome mounting resistance. Nanotechnology stands out, with innovations such as metallic (silver, gold, zinc), polymeric, dentrimeric, liposomal, and lipid-based nanoparticles. These can disrupt biofilm structure, enhance antibiotic penetration, deliver drugs in a controlled fashion, and even act as quorum-sensing inhibitors [24]. For instance, integrating nanoparticles with compassionate carriers such as mesoporous silica is showing promise in penetrating and dismantling biofilm defense (*see* **Note 4**) (Mayorga-Ramos).

Quorum-sensing (QS) inhibition, which disrupts microbial communication that governs biofilm formation, is also gaining traction. Natural compounds, such as ajoene, and enzymes from specific bacteria have been shown to degrade or interfere with QS molecules, suppressing biofilm formation across both Gram-negative and Gram-positive species.

3 Conclusion

Microbial biofilms represent one of the most resilient and complex forms of microbial life, playing critical roles in environmental processes, industrial applications, and clinical challenges. Throughout this book, we have explored the fundamental biology of biofilm formation, the mechanisms of microbial interactions within the EPS matrix, and the significant impacts of biofilms in diverse settings. Central to advancing biofilm research and developing effective control strategies is the ability to accurately identify, characterize, and monitor biofilms in situ and in real time.

State-of-the-art identification methods—including advanced microscopy techniques, molecular and biochemical assays, omics-based approaches, and novel label-free sensors—have revolutionized our understanding of biofilm structure, function, and heterogeneity. These tools have enabled researchers to dissect biofilm dynamics at multiple scales, from single-cell behavior to community-level interactions, while providing critical insights into biofilm resilience and antimicrobial resistance.

Despite significant progress, challenges remain in detecting biofilms in complex environments, such as within medical implants, industrial piping systems, or natural ecosystems, where biofilm presence is often hidden or difficult to quantify. Ongoing developments in microfluidics, real-time impedance monitoring, and computational image analysis offer promising avenues to overcome these limitations, providing more sensitive, specific, and noninvasive solutions.

Looking ahead, the integration of interdisciplinary approaches combining microbiology, engineering, bioinformatics, and materials science will be key to further unravelling biofilm complexity. This will not only improve identification and monitoring but also enable the development of targeted strategies for prevention, disruption, and eradication of harmful biofilms, while harnessing beneficial biofilms in bioremediation and industrial biotechnology.

This work serves as a foundational resource for researchers, clinicians, and industry professionals seeking to deepen their understanding of biofilm identification methods and their applications. By advancing these tools and approaches, we move closer to effective biofilm management solutions that address critical challenges in medicine, industry, and the environment.

4 Notes

1. Common Problems

- Weak or no biofilm formation
- High background staining
- Variability between replicates
- False positives or negatives

Main Causes

- Incorrect media composition (e.g., glucose concentration)
- Improper dye concentration or expired reagents
- Excessive or insufficient washing
- Inconsistent inoculum density
- Edge effects in microtiter plates

Key Solutions

- Standardize inoculum (0.5 McFarland)
- Use TSB + 1% glucose where required
- Wash gently but thoroughly with PBS
- Avoid outer wells in 96-well plates
- Include positive controls (e.g., *Pseudomonas aeruginosa*) and negative controls
- Perform assays in triplicate

2. Common Problems

    - Biofilm detachment during preparation
    - Poor staining or weak fluorescence
    - Structural distortion (SEM)

    Main Causes

    - Harsh washing
    - Inadequate fixation
    - Expired or improperly handled dyes
    - Rapid dehydration steps

    Key Solutions

    - Fix samples before staining
    - Handle gently during washing
    - Prepare fresh staining reagents
    - Use gradual dehydration for SEM

3. Common Problems

    - No gene amplification
    - Nonspecific bands

    Main Causes

    - Poor DNA quality
    - PCR inhibitors
    - Incorrect annealing temperature
    - Primer issues

    Key Solutions

    - Verify DNA purity (A260/A280 $\approx$ 1.8)
    - Use positive control strains (e.g., *ica* genes in *Staphylococcus epidermidis*)
    - Optimize annealing temperature
    - Use gradient PCR if needed

4. Common Problems

    - Low metabolic signal despite visible biofilm

    Main Causes

    - Dormant cells
    - Reagent degradation
    - Incorrect incubation time

Key Solutions

- Use freshly prepared reagents
- Protect from light
- Optimize incubation time (2–4 h)
- Maintain consistent incubation temperature (usually 37 °C)
- Always include positive, negative, and blank controls
- Perform experiments in triplicate
- Record all experimental parameters carefully
- Repeat experiments independently to confirm reproducibility

Competing Interest

The author declares that there are no known financial, personal, academic, or professional conflicts of interest that could have appeared to influence the work reported in this chapter.

References

1. Costerton JW, Lewandowski Z, Caldwell DE, Korber DR, Lappin-Scott HM (1995) Microbial biofilms. Ann Rev Microbiol 49(1): 711–745. https://doi.org/10.1146/annurev.mi.49.100195.003431
2. Flemming H-C, Wingender J (2010) The biofilm matrix. Nat Rev Microbiol 8(9):623–633. https://doi.org/10.1038/nrmicro2415
3. Donlan RM, Costerton JW (2002) Biofilms: survival mechanisms of clinically relevant microorganisms. Clin Microbiol Rev 15(2): 167–193. https://doi.org/10.1128/CMR.15.2.167-193.2002
4. Hall-Stoodley L, Costerton JW, Stoodley P (2004) Bacterial biofilms: from the natural environment to infectious diseases. Nat Rev Microbiol 2(2):95–108. https://doi.org/10.1038/nrmicro821
5. Battin TJ, Kaplan LA, Newbold JD, Hansen CM (2007) Contributions of microbial biofilms to ecosystem processes in stream mesocosms. Nature 426(6965):439–442. https://doi.org/10.1038/nature02002
6. Liu HY, Prentice EL, Webber MA (2024) Mechanisms of antimicrobial resistance in biofilms. NPJ Antimicrob Resist 2(1):27. https://doi.org/10.1038/s44259-024-00046-3
7. Knisz J, Eckert R, Gieg LM, Koerdt A, Lee JS, Silva ER, Skovhus TL, An Stepec BA, Wade SA (2023) Microbiologically influenced corrosion-more than just microorganisms. FEMS Microbiol Rev 47(5):fuad041. https://doi.org/10.1093/femsre/fuad041
8. Pereira MO et al (2022) Quorum sensing in biofilms: relevance to clinical infections. Front Cell Infect Microbiol 12:868282. https://doi.org/10.3389/fcimb.2022.868282
9. Wang X, Liu M, Yu C, Li J, Zhou X (2023) Biofilm formation: mechanistic insights and therapeutic targets. Mol Biomed 4(1):49. https://doi.org/10.1186/s43556-023-00164-w
10. Sharma S, Mohler J, Mahajan SD, Schwartz SA, Bruggemann L, Aalinkeel R (2023) Microbial biofilm: a review on formation, infection, antibiotic resistance, control measures, and innovative treatment. Microorganisms 11(6): 1614. https://doi.org/10.3390/microorganisms11061614
11. Malygina A, Balkin A, Polyakova E, Stefanov S, Potekhin A, Gogoleva N (2023) Taxonomic diversity of the microbial biofilms collected along the thermal streams on Kunashir Island. Ecologies 4(1):106–123. https://doi.org/10.3390/ecologies4010009
12. Fröls S (2013) Archaeal biofilms: widespread and complex. Biochem Soc Trans 41(1): 393–398. https://doi.org/10.1042/bst20120304
13. Calderón K, González-Martínez A, Gómez-Silván C, Osorio F, Rodelas B, González-López J (2013) Archaeal diversity in biofilm technologies applied to treat urban and

industrial wastewater: recent advances and future prospects. Int J Mol Sci 14(9): 18572–18598. https://doi.org/10.3390/ijms140918572

14. Parrilli E, Tutino ML, Marino G (2022) Biofilm as an adaptation strategy to extreme conditions. Rend Fis Acc Lincei 33:527–536. https://doi.org/10.1007/s12210-022-01083-8
15. Yin W, Wang Y, Liu L, He J (2019) Biofilms: The Microbial "Protective Clothing" in Extreme Environments. Int J Mol Sci 20(14): 3423. https://doi.org/10.3390/ijms20143423. PMID: 31336824; PMCID: PMC6679078
16. Silva NBS, Marques LA, Röder DDB (2021) Diagnosis of biofilm infections: current methods used, challenges and perspectives for the future. J Appl Microbiol 131:2148–2160. https://doi.org/10.1111/jam.15049
17. Wang C, Su Y, Shahriar SMS, Li Y, Xie J (2024) Emerging strategies for treating medical device and wound-associated biofilm infections. Microb Biotechnol 17(10):e70035. https://doi.org/10.1111/1751-7915.70035
18. Huang Y, Chen P, Cao H, Zhou Z, Xu T (2024) Characterization of *Pseudomonas aeruginosa* isolated from bovine mastitis in northern Jiangsu Province and correlation to drug resistance and biofilm formability. Animals 14(22):3290. https://doi.org/10.3390/ani14223290
19. Xie Y, Liu H, Teng Z, Ma J, Liu G (2025) Nanomaterial-enabled anti-biofilm strategies: new opportunities for treatment of bacterial infections. J Nanoscale 17(10). https://doi.org/10.1039/D4NR04774E
20. Abbas SH, Khan S, Shah M, Aslam J, Nawaz H, Ilyas N, Gamaryani A, Afridi SQ, Khan I, Shah B, Shah K, Rashid A, Khan D, Khan S (2024) Public health threats posed by biofilms and innovative strategies for their control. Discoveries (Craiova) 12(4):e197. https://doi.org/10.15190/d.2024.16
21. Singh A, Amod A, Pandey P, Bose P, Pingali MS, Shivalkar S, Varadwaj PK, Sahoo AK, Samanta SK (2022) Bacterial biofilm infections, their resistance to antibiotics therapy and current treatment strategies. Biomed Mater 17(2). https://doi.org/10.1088/1748-605x/ac50f6
22. Mishra R, Panda AK, De Mandal S, Shakeel M, Bisht SS, Khan J (2020) Natural anti-biofilm agents: strategies to control biofilm-forming pathogens. Front Microbiol 11. https://doi.org/10.3389/fmicb.2020.566325
23. Mayorga-Ramos A, Carrera-Pacheco SE, Barba-Ostria C, Guamán LP (2024) Bacteriophage-mediated approaches for biofilm control. Front Cell Infect Microbiol 14. https://doi.org/10.3389/fcimb.2024.1428637
24. Iaconis A, De Plano LM, Caccamo A, Franco D, Conoci S (2024) Anti-biofilm strategies: a focused review on innovative approaches. Microorganisms 12(4):639. https://doi.org/10.3390/microorganisms12040639

Part II

Microscopy-Based Methods

Chapter 2

Microscopy-Based Methods for Visualizing Biofilm Architecture and Structure: Light Microscopy

Rajalakshmi M, Srinivasan P, and Amritha Sathya P

Abstract

Light microscopy remains a mainstay technique for visualizing biofilm architecture, providing versatile modalities for the observation, quantification, and analysis of biofilms under near-physiological conditions. This chapter details comprehensive protocols including biofilm cultivation, sample fixation, staining, and imaging techniques involving bright-field, phase contrast, differential interference contrast (DIC), and fluorescence microscopy. It emphasizes practical guidance for image acquisition and quantitative morphometric analysis. Extensive notes on troubleshooting, artifact management, and reproducibility enhancement support researchers at all expertise levels. The protocols enable detailed characterization of biofilm biomass, cellular viability, and matrix heterogeneity, facilitating insights into biofilm biology and function.

Key words Biofilm, Light microscopy, Bright-field, Phase contrast, Differential interference contrast, Fluorescence microscopy, Staining, Architecture, Quantitative analysis

1 Introduction

Biofilms are structured microbial communities embedded within a self-produced matrix of extracellular polymeric substances (EPS), adhering to biotic or abiotic surfaces and representing a predominant mode of microbial life in natural and artificial environments [1]. They exhibit altered phenotypes compared to planktonic cells that include increased resistance to antimicrobials and environmental stresses, posing significant challenges in clinical, industrial, and environmental contexts [6, 8]. Detailed visualization of biofilm architecture, including spatial organization, viability patterns, and matrix distribution, is critical to understanding biofilm physiology, pathogenesis, ecological function, and for devising strategies for control or exploitation.

Light microscopy provides essential techniques for biofilm imaging, offering a balance of accessibility, real-time imaging capability, and compatibility with various sample preparations and

Abirami K (ed.), *Microbial Biofilm Research: Comprehensive Protocols and Best Practices*, Springer Protocols Handbooks,
https://doi.org/10.1007/978-1-0716-5535-1_2,

staining protocols [2]. Bright-field microscopy, augmented by stains such as crystal violet, enables visualization of overall biomass and biofilm mass distribution [3]. Phase contrast and differential interference contrast (DIC) microscopy facilitate label-free, high-contrast imaging of live or fixed biofilms by converting refractive index differences into image contrast [9]. Fluorescence microscopy applies specific fluorophores or genetically encoded fluorescent markers for high-sensitivity visualization of live/dead cells, polysaccharides, and microbial populations, allowing multiplexed studies of biofilm composition and function [5].

While the resolution of conventional light microscopy is limited by the diffraction of light to approximately 200 nm, it remains invaluable for mesoscale examination of biofilm structures under conditions that preserve hydration and cellular viability, unlike electron microscopy, which demands extensive sample processing [10]. Recent advancements in optics, digital imaging, and image analysis software significantly enhance the ability to quantify biofilm structure and dynamics, aiding reproducibility and rigor.

This chapter presents detailed, stepwise protocols to cultivate biofilms, prepare samples, apply appropriate staining, and acquire high-quality images across commonly used light microscopy modalities. It includes practical tips, troubleshooting guides, and quantitative image analysis methods to equip researchers, irrespective of prior light microscopy experience, with tools to rigorously characterize biofilm architecture and biology.

2 Materials

The choice and preparation of buffers, solutions, media, and specialized equipment critically affect the quality and reproducibility of biofilm imaging experiments using light microscopy techniques. This section provides detailed compositions, preparation protocols, and practical advice for all essential materials required to reliably cultivate, stain, and image biofilms on microscopic surfaces.

2.1 Buffers and Solutions

2.1.1 Phosphate-Buffered Saline (PBS)

Phosphate-buffered saline (PBS) is an isotonic, non-toxic buffer solution widely used for washing, diluting, and maintaining physiological pH during biofilm staining and microscopy preparation. Its composition mimics physiological salt concentrations and pH to preserve cellular and matrix integrity without causing osmotic stress or chemical damage to biofilm structures.

Recipe and Preparation

- Weigh and dissolve the following in ~800 mL distilled, deionized water:
  - Sodium chloride (NaCl): 8.0 g.
  - Potassium chloride (KCl): 0.2 g.

  - Disodium hydrogen phosphate (Na_2HPO_4): 1.44 g.
  - Potassium dihydrogen phosphate (KH_2PO_4): 0.24 g.

- After complete dissolution, adjust pH carefully to 7.4 using dilute hydrochloric acid (HCl) or sodium hydroxide (NaOH).
- Bring the final volume up to 1 liter with distilled water.
- Sterilize by autoclaving at 121 °C for 20 min or filter sterilize (0.22 μm pore size).
- Store at 4 °C, protected from light and contamination, ideally prepare fresh monthly to avoid microbial growth or precipitation [6].

Practical Notes

- Ensuring correct pH is critical, as pH shifts can affect EPS structure and staining.
- Use high purity reagents and deionized water to prevent precipitates or contaminants.
- PBS is used to gently wash biofilms without causing mechanical disruption to attached communities.

2.1.2 Crystal Violet Stain (0.1% w/v)

Crystal violet is a basic dye with strong affinity for negatively charged components (polysaccharides, nucleic acids) within biofilm biomass. It is classic for staining total biofilm mass in bright-field microscopy and colorimetric biomass quantification assays.

Preparation Protocol

- Weigh precisely 0.1 g of crystal violet powder, preferably analytical grade.
- Dissolve slowly in 100 mL distilled water or in PBS to aid compatibility.
- Stir gently for 20–30 min until fully dissolved.
- Filter through a 0.22 μm sterile filter to remove undissolved particles or contaminants that may cause background noise or uneven staining [13].
- Store in amber or opaque bottles at room temperature but protected from direct sunlight to prevent dye degradation.

Usage Notes

- Dilutions other than 0.1% are sometimes used; 0.1% is optimal for structural imaging without excessive background staining.
- Prepare fresh batches monthly or as needed to ensure staining consistency.

2.1.3 Safranin Stain (0.5% w/v)

Safranin is a positively charged dye that preferentially binds acidic polysaccharides in the EPS matrix, useful to complement crystal violet staining and to highlight biofilm extracellular matrix components in bright-field microscopy.

Preparation Protocol

- Accurately weigh 0.5 g of safranin powder.
- Add to 100 mL distilled water at room temperature.
- Stir continuously until fully dissolved, which may take up to 30 min depending on grain size and water temperature.
- Filter sterilizes using 0.22 μm filters to remove residue and particulate impurities.
- Store in amber bottles shielded from direct light, at room temperature or refrigerated.

Practical Considerations

- Safranin is sensitive to light; exposure can diminish staining efficacy.
- Used as a counterstain following crystal violet or as a separate matrix indicator.

2.1.4 Live/Dead Staining Solution

To assess viability within biofilms, fluorescent stains such as SYTO9 and propidium iodide (PI) are combined in commercial kits to distinguish live (membrane-intact) from dead (membrane-compromised) cells. SYTO9 permeates all cells staining nucleic acids green, while PI penetrates only damaged membranes producing red fluorescence.

Preparation and Storage

- Use commercial kits following manufacturer instructions for dilution and handling.
- Protect from light and use freshly prepared solutions or store aliquots at −20 °C if stable.
- Optimize staining duration (usually 15 mins) and concentrations empirically for specific biofilms [5].

Application Guidelines

- Incubate biofilms with the dye mix in darkness to preserve fluorescence.
- Rapid processing post-staining maintains dye integrity and reduces photobleaching.

2.1.5 Mounting Medium

Mounting medium is required to preserve stained biofilms on slides, prevent dehydration, and protect fluorophore integrity in fluorescence imaging.

Common Types

- *Glycerol/PBS mixture (1:1)*: Provides refractive index matching and preserves sample hydration.
- *Commercial antifade reagents*: Incorporate antioxidants that mitigate fluorophore photobleaching during prolonged microscopy sessions.

Preparation

- Mix glycerol and PBS freshly or purchase ready-to-use antifade mounting solutions.
- Filter sterilizes if necessary.
- Apply sparingly under coverslips during slide preparation.

2.1.6 Growth Media

Culture media composition significantly affects biofilm development, influencing biomass, matrix production, and cell physiology.

Luria-Bertani (LB) Broth
A nutrient-rich, general-purpose medium widely used for culturing bacteria.

Preparation

- Dissolve 10 g tryptone, 5 g yeast extract, and 10 g NaCl per liter in distilled water.
- Adjust pH to ~7.0 if needed.
- Sterilize by autoclaving (121 °C, 20 min).
- Store in sterile containers at room temperature.

Usage

1. Suitable for a broad range of bacteria such as *Escherichia coli* and *Pseudomonas* species.
2. Supports rapid growth resulting in rapid biofilm formation timelines.

2.1.7 Minimal Media

Defined media contain specific nutrients tailored to the metabolic requirements of particular microorganisms or experimental designs, enabling controlled studies of biofilm physiology.

Preparation

- Recipes vary, commonly containing salts, a carbon source (e.g., glucose), nitrogen source, vitamins, and minerals.
- Prepare stock solutions of components and mix fresh; sterilize by autoclaving or filtration.
- Examples include M9 minimal medium or defined synthetic biofilm media.

Applications

- Useful for studies examining specific nutrient effects or producing EPS component variation.
- Enables reproducible biofilm architectures suited for advanced imaging analysis.

2.2 Equipment

2.2.1 Light Microscope

- Equipped with a range of objectives: typically, 10x and 40x dry objectives for overview biofilm imaging, and a 100x oil immersion lens for high-resolution cellular details.
- Bright-field, phase contrast, and differential interference contrast (DIC) modalities ideally available on the same microscope for multi-technique versatility.

Optical Modules

- *Phase Contrast*: Essential for label-free imaging of live or unstained biofilms, consisting of a specialized condenser annulus and matching objectives to convert phase shifts into contrast.
- *DIC (Nomarski) Optics*: Include Wollaston prisms and polarizers to generate detailed surface topology images with pseudo-3D contrast, invaluable for morphological biofilm studies.

2.2.2 Fluorescence Microscope

- Equipped with filter sets for *FITC* (green), *TRITC* or *Texas Red* (red), and *DAPI* (blue) excitation/emission spectra, enabling multispectral imaging of stained biofilm components.
- Ideally includes a high-sensitivity camera and software for z-stack (3D) acquisition.

2.2.3 Sample Holders

- Cleaned and sterile glass microscope slides and coverslips sized 22 × 30 mm commonly used; microtiter plates can be adapted for high-throughput assays with transparent bottoms suitable for microscopy.

2.2.4 Incubator

- Maintained at temperatures from 25 to 37 °C as per microbial growth requirements, with controlled humidity to prevent evaporation and maintain biofilm integrity during incubation.

2.2.5 Liquid Handling

- Pipettes spanning volumes from 0.1 μL to 10 mL with sterile tips for precise reagent and culture handling.

2.2.6 Imaging Hardware and Software

- High-quality *digital camera* or *CCD/sCMOS sensors* attached to microscopes for image capture.
- Computers loaded with image processing platforms such as *ImageJ*, *Fiji*, or proprietary microscope software for image analysis, quantification, and archiving.

Practical Considerations and Tips

- Use fresh reagents and sterile techniques to prevent contamination or reagent degradation.
- Validate buffer pH and sterility regularly.
- Choose mounting media compatible with fluorophores in use to avoid quenching signals.
- Regular microscope maintenance (cleaning lenses, alignment) optimizes image quality.
- Adapt media and incubation times to your specific microbial strains and experimental needs; pilot studies are encouraged.
- Document reagent lot numbers and preparation dates for reproducibility.

3 Methods

3.1 Biofilm Cultivation

Biofilm formation is a complex and dynamic process where microorganisms attach to a surface, proliferate, and produce extracellular polymeric substances (EPS) that anchor the community. Proper cultivation of biofilms under controlled laboratory conditions is critical to obtaining reproducible structures suitable for microscopy and downstream analysis. Various methods are employed depending on the biological question and sample type, with static culture on abiotic surfaces and flow-cell systems being the most prevalent.

3.1.1 Preparation of Inoculum

Before biofilm cultivation, a uniform and healthy inoculum is required to ensure reproducible initiation of biofilm growth.

- Start by streaking the microbial strain of interest on a suitable agar plate (e.g., LB agar for many bacteria). Incubate at optimal temperature (e.g., 37 °C) overnight.
- Select a single colony and inoculate into liquid medium such as Luria-Bertani (LB) broth or strain-specific media, depending on preferences and growth requirements.
- Incubate with agitation (e.g., 150–200 rpm) at the organism's optimal temperature (commonly 30–37 °C) until reaching

stationary phase, typically 12–18 h. Stationary phase cultures generally exhibit enhanced biofilm-forming capacity [1, 6].

- Measure optical density at 600 nm (OD600) with a spectrophotometer. Use this data to dilute the culture into fresh medium to a standard inoculum concentration, typically OD600 = 0.1, corresponding to approximately 10^6 cells/mL, to standardize biofilm formation conditions (*see* **Note 1**).

3.1.2 Static Biofilm Culture

Static culture systems constitute a straightforward method for biofilm formation, suitable for free-floating species, but with limitations in simulating fluid dynamics.

- Transfer standardized inoculum (1–5 mL) into sterile polystyrene multiwell plates (e.g., 24- or 96-well plates), petri dishes, or glass tubes containing sterile abiotic surfaces such as glass coverslips or microscope slides. Coverslips provide a transparent surface compatible with light microscopy [2].
- For surface biofilm studies, place sterile glass coverslips vertically or horizontally in wells to allow attachment on both sides; adjust volume to submerge the surface without overflow.
- Incubate plates statically at appropriate temperatures (25–37 °C) for defined durations (24, 48, or 72 h are common). Longer incubation promotes biofilm maturation with thick extracellular matrix accumulation [3].
- Avoid disturbance during incubation to prevent biofilm detachment. Humidity chambers can minimize evaporation in open vessels (*see* **Note 2**).

3.1.3 Flow Cell Biofilm Culture

Dynamic biofilm cultivation systems more closely mimic natural environments found in medical and industrial contexts, providing continuous nutrient supply and shear stress.

- Assemble flow cells with transparent surfaces (usually glass) designed for microscopic imaging; commercial or custom-built chambers are available [4].
- Sterilize components thoroughly by autoclaving or chemical sterilant.
- Inoculate the flow cell by pumping cell suspension through the system for 1–2 h to allow initial adhesion.
- Switch to continuous fresh medium flow (0.1–1 mL/min) using peristaltic pumps to sustain biofilm growth under defined shear forces [5].
- Incubate under constant temperature control. Flow rate and nutrient composition can be adjusted to study shear effects, nutrient gradients, or biofilm resilience.

- This method permits in situ live imaging with minimal disturbance (*see* **Note 3**).

3.1.4 Biofilm Harvesting Considerations

- Harvest biofilms gently to preserve structural integrity; avoid scraping or vigorous pipetting prior to fixation.
- If necessary, downstream biofilm dispersal or extraction should be described separately to prevent artefacts in microscopy (*see* **Note 4**).

3.2 Biofilm Fixation

Fixation preserves biofilm architecture and cellular morphology, stabilizing structural components for staining and imaging.

3.2.1 Washing

- Post incubation, remove planktonic cells by gently washing biofilms with sterile phosphate-buffered saline (PBS).
- Use gentle pipetting along edges or immerse substrata in PBS baths to minimize biofilm detachment [6].
- Repeat 2–3 times until supernatant is clear.

3.2.2 Fixation with Paraformaldehyde (PFA)

- Prepare 4% PFA freshly by dissolving paraformaldehyde powder in PBS at 60 °C with pH adjustment, cooling to room temperature before use.
- Immerse biofilm samples in fixation solution for 20–30 min at room temperature.
- PFA crosslinks amine groups in proteins, stabilizing cell and matrix structures while preserving epitopes for antibody or fluorescent staining [7].
- After fixation, wash samples twice with PBS to remove residual fixative, which can cause background fluorescence and chemical toxicity.

3.2.3 Alternative Fixatives

- Cold methanol fixation (−20 °C for 10–15 min) permeabilizes membranes beneficially for nucleic acid stains but can cause dehydration artifacts and is less suitable for studying matrix polysaccharides [8].
- Glutaraldehyde fixation offers stronger crosslinking but is generally reserved for electron microscopy due to auto fluorescence and decreased antigenicity.

3.2.4 Storage

- Fixed samples can be stored at 4 °C in PBS, protected from light, for up to 24 h before staining. Longer storage may compromise fluorescence and biofilm integrity (*see* **Note 5**).

3.3 Staining Protocols

3.3.1 Crystal Violet Staining (Total Biomass)

- Crystal violet is a cationic dye binding strongly to negatively charged surface polymers and DNA, staining both cells and EPS.
- Prepare fresh 0.1% crystal violet solution and submerge samples for 10–15 min at room temperature.
- Wash gently with distilled water to remove unbound dye until runoff is colorless, avoiding harsh rinsing that may detach biofilm layers.
- Air dry completely to enhance contrast in bright-field imaging [9].
- Crystal violet can also be quantified spectrophotometrically after solubilizing the dye in ethanol, though this chapter focuses on microscopy applications.

3.3.2 Safranin Staining (EPS Polysaccharides)

- Safranin stains acidic polysaccharides in the biofilm matrix.
- Incubate fixed biofilms in 0.5% aqueous safranin for 5 min.
- Rinse gently with distilled water and air dry.
- Use in combination with crystal violet to discriminate matrix from cellular components, aiding structural evaluation [10].

3.3.3 Live/Dead Fluorescent Staining

- Use commercially available SYTO9/propidium iodide kits that distinguish live (green fluorescence) from dead (red fluorescence) cells based on membrane integrity.
- Dilute dyes as recommended and incubate samples in the dark for 15 min.
- Wash samples gently with PBS to remove excess stain.
- Immediate imaging is critical to avoid signal degradation and photobleaching [11].
- Fluorescence microscopy enables quantification of viability distributions within heterogeneous biofilms.

3.3.4 Additional Staining Options

- Lectins (e.g., Concanavalin A-FITC) to visualize specific EPS components.
- Nucleic acid dyes (DAPI, SYBR Green) for total cell visualization.
- Membrane stains (FM4-64) to outline bacterial cell boundaries.
- Protocol optimization of stain concentration and incubation times is recommended depending on the strain and experimental aims.

3.4 Mounting and Slide Preparation

- After staining, apply ~20 μL mounting medium (e.g., glycerol/PBS 1:1) on biofilm-coated areas.
- Slowly place a clean coverslip to avoid air bubbles, which interfere with focusing and image clarity.
- Seal coverslip edges with clear nail polish or mounting medium to prevent drying and sample movement during imaging.
- For fluorescence samples, protect from light during storage (4 °C) and imaging to preserve fluorescence [12, 14].

3.5 Microscopy Imaging

3.5.1 Bright-Field Microscopy

- Starting at low magnifications (10x, 20x) locate uniform biofilm regions suitable for analysis [15, 16].
- Increase magnification to 40x or 100x oil immersion for detailed cellular structure visualization.
- Adjust condenser position and iris diaphragms to enhance contrast while avoiding glare.
- Capture multiple areas per sample and across biological replicates for robust data (Fig. 1).

3.5.2 Phase Contrast Microscopy

- Use phase contrast optics with matched phase annulus and condenser ring for enhanced visualization of unstained, transparent biofilms.

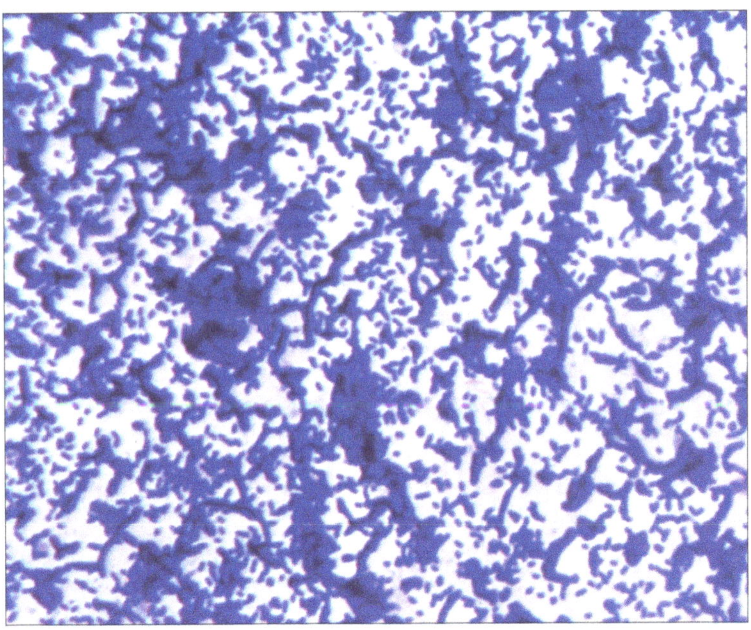

Fig. 1 Crystal violet-stained *Pseudomonas aeruginosa* biofilm on a glass coverslip. Bright-field microscopy image taken at 40× magnification showing dense micro colonies embedded in extracellular polymeric substances. Scale bar = 50 μm

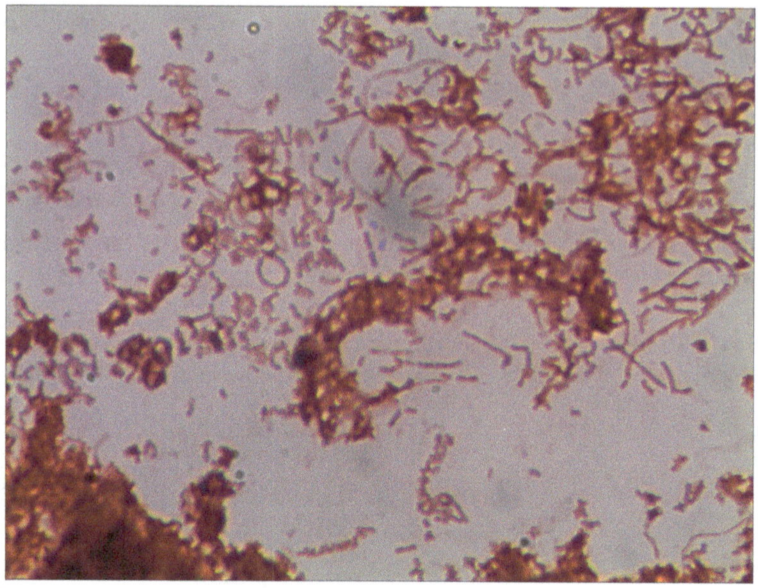

Fig. 2 Phase contrast microscopy of live *Escherichia coli* biofilm grown under static conditions. Cell clusters and matrix appear as bright structures with clear boundaries. Magnification 100×, oil immersion

- Ensure precise alignment and calibration for sharp halo effects emphasizing cell boundaries.
- Ideal for live biofilms where fixation or staining might alter biological condition (Fig. 2).

3.5.3 Differential Interference Contrast (DIC) Microscopy

- Install Wollaston prisms and appropriate polarizers for DIC imaging.
- Adjust shear and bias retardation to generate pseudo-3D images that reveal surface topography and matrix features.
- Use for detailed morphology of extracellular matrix and cell-matrix interfaces.

3.5.4 Fluorescence Microscopy

- Select appropriate excitation and emission filters (e.g., FITC for SYTO9, TRITC for propidium iodide).
- Balance illumination intensity and exposure time to maximize signal while minimizing photobleaching.
- Perform z-stack imaging for three-dimensional reconstructions if required (*see* **Notes 6**).
- Use controls stained with individual dyes and unstained samples for signal specificity verification [16] (Fig. 3).

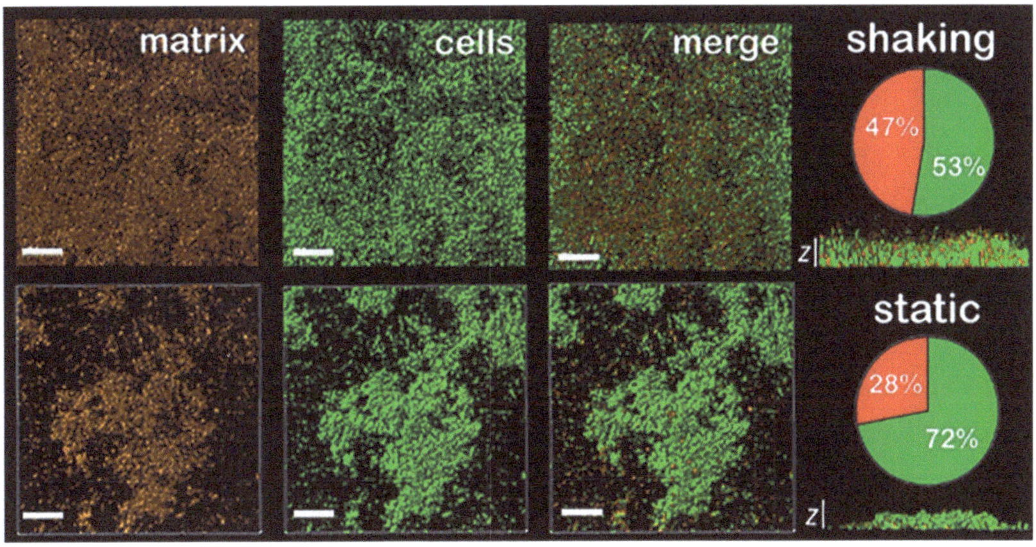

Fig. 3 Differential interference contrast (DIC) image of a mature biofilm surface topology. The image reveals three-dimensional surface morphology and matrix distribution in *Pseudomonas fluorescens*. Magnification 60×

3.6 Image Acquisition and Analysis

- Use high-sensitivity CCD or sCMOS cameras attached to the microscope for image capture, ensuring consistent illumination across experiments.
- Employ image analysis platforms such as ImageJ or Fiji for data processing: background subtraction, thresholding, segmentation of biofilm biomass, and viable/non-viable populations.
- Quantify biofilm thickness from z-stack data by analyzing fluorescence or bright-field intensity profiles.
- Calculate surface coverage, total biomass area, and live/dead cell ratios.
- Export raw and processed data with scale calibration for downstream statistical analysis and publication.

4 Notes

1. Adjust inoculum concentration and growth conditions to optimize biofilm formation for specific microbial strains, as biofilm robustness can vary considerably.
2. Minimize evaporation in static cultures by sealing plates or incubating in humidity-controlled chambers.
3. Flow cell systems require careful sterile technique and calibration to prevent contamination and ensure stable flow.
4. Consider biofilm heterogeneity when selecting areas for imaging to avoid biased quantification.

5. Storage of fixed samples longer than 24 h may decrease staining quality; perform imaging promptly.
6. Photobleaching is a significant concern in fluorescence microscopy; minimize light exposure and use antifade mounting agents.

(a) *Variability in Biofilm Formation and Inoculum Preparation*

Biofilm formation is inherently heterogeneous due to strain-specific factors, environmental conditions, and stochastic biological processes. Ensuring an accurately standardized inoculum is critical. Cultures should ideally be harvested during late exponential or early stationary phase to maximize expression of adhesion molecules and EPS synthesis enzymes, which are key drivers of robust biofilm development. Using older cultures can result in decreased viability and variable attachment, degrading reproducibility. Prepare inocula freshly and measure optical density carefully using a calibrated spectrophotometer to avoid discrepancies. Dilution steps must be precise, and working with biological triplicates is recommended to account for intrinsic variability. Additionally, inoculum passage history and storage conditions affect phenotypic stability and biofilm-forming capacity; minimize freeze-thaw cycles and ensure consistent revival practices.

(b) *Control of Environmental Parameters During Static Culture*

Static biofilm cultivation is sensitive to local microenvironmental fluctuations such as oxygen gradients, nutrient depletion, and evaporation. To minimize evaporation, which can concentrate salts and alter osmolarity, cultures should be incubated in sealed humidified chambers or plates sealed with breathable membranes that maintain sterility while limiting moisture loss. Consider incubator humidity and temperature stability, as fluctuations can impact biofilm architecture. Maintain incubators according to manufacturer specifications and calibrate periodically. If possible, monitor plate humidity using sensors, especially for long incubation periods exceeding 48 h.

(c) *Flow Cell System Optimization and Contamination Prevention*

Flow cells simulate in vivo and industrial environments by providing controlled shear stress and constant nutrient flow, impacting biofilm morphology profoundly. Precise assembly of flow systems under sterile conditions is essential; use aseptic techniques when assembling tubing sets and flow chambers. To prevent contamination, sterilize individual components with autoclaving or chemical disinfectants compatible with materials (e.g., ethanol, bleach followed by extensive rinsing). Between experiments, clean flow cells thoroughly by chemical and mechanical means to remove residual biofilms that can serve as inoculum sources or obstruct flow. Pre-flush systems with sterile medium and verify flow rates with calibrated instruments. Transfer biofilm cultures into sterile reservoirs and tubing using sterile connectors or syringes. Ensure bubble traps are installed to avoid flow disruption which can cause biofilm sloughing.

(d) *Minimizing Biofilm Disturbance During Washing*

The fragile nature of biofilms, especially in early or thin stages, requires gentle washing procedures to avoid inadvertent removal of attached biomass. It is recommended to pipette washing buffer slowly along the edges of substrata rather than directly onto biofilms. Alternatives include immersion of coverslips in large volumes of washing buffer with gentle rocking. Using wide-bore pipette tips may reduce shear forces during washing steps. Avoid multiple rapid washes which increase the risk of detachment.

(e) *Fixation Artifacts and Their Management*

Chemical fixation stabilizes biofilm architecture but can introduce artifacts if not carefully performed. Paraformaldehyde (PFA) is preferred for preserving proteinaceous structures with minimal shrinkage; however, poor fixation due to insufficient concentration, improper pH, or old reagents may lead to sample degradation or loss of antigenicity critical for immunolabeling. Methanol fixation is rapid and preserves nucleic acid structures but often causes dehydration and EPS collapse affecting biofilm ultrastructure. Use methanol fixation sparingly or when targeting specific components. Test fixation approaches on pilot samples prior to large experiments. Always prepare fresh PFA solutions, verify pH between 7.2 and 7.4, and ensure samples are fixed for adequate durations (20–30 min) at room temperature or at 4 °C for higher antigen preservation.

(f) *Optimization of Staining Protocols*

Staining parameters must be finely tuned to the biofilm species and experimental objectives. Overstaining can create high background signals impeding contrast, while insufficient staining results in poor visualization and loss of detail. Titrate stain concentrations and incubation times during preliminary experiments. Include unstained controls and single-stain controls to discern true signal from autofluorescence, quenching, or spillover. Use fresh staining solutions, as dyes degrade over time especially with light exposure. For fluorescence stains, avoid direct light during staining and storage. Protect samples in amber containers or wrapped in foil, and use antifade mounting media to preserve fluorescence during imaging.

(g) *Avoidance of Mounting Artifacts*

Air bubbles trapped beneath coverslips distort light paths causing uneven illumination and focus loss. Place coverslips slowly at an angle rather than directly perpendicular to minimize bubble formation. Seal edges with clear nail polish or mounting medium to prevent slow drying and sample movement. For multi-hour or multi-day imaging relative humidity control or enclosed chambers are desirable to maintain sample hydration and fluorophore stability.

(h) *Microscope Alignment and Maintenance*

Proper optical alignment is paramount for acquiring high-quality images. Regularly clean objective lenses, phase contrast annuli, Wollaston prisms (for DIC), condensers, and filter cubes using appropriate lens paper and solvents (e.g., ethanol or lens cleaner). Check phase contrast conditions by viewing reference beads or standard biological samples to adjust phase plates and condenser diaphragms for optimal contrast and resolution. Verify excitation source intensity uniformity and alignment for fluorescence microscopy. Malfunctioning filters or lamps/detectors degrade image quality and quantification reliability. Document microscope settings such as light intensity, exposure time, and camera gain for reproducibility.

(i) *Standardization of Image Acquisition*

Capture images from multiple, randomly selected fields of view within each sample to address biofilm heterogeneity. Uniform microscope settings across all samples in an experiment maintain data comparability. Use neutral density filters or adjustable lamp power to manage illumination and prevent saturation. Perform image acquisition in raw formats when possible, enabling post-acquisition adjustments without data loss. Utilize stage coordinates or automated tiling to systematically image entire substrate surfaces for comprehensive biofilm mapping.

(j) *Quantitative Image Analysis Pitfalls and Best Practices*

Image processing is susceptible to user bias; hence apply thresholding algorithms consistently and validate segmentation with raw images. Calibrate pixel dimensions using stage micrometers for accurate spatial measurements. Clearly report thresholds and analysis parameters in publications. Analyzing biological replicates statistically strengthens conclusions. Software such as ImageJ/Fiji offers robust plugins and macros for batch processing, integral for large datasets. Avoid over-interpretation of artifacts or staining inconsistencies as biological phenomena by including proper controls and complementary assays where possible.

(k) *Photo Bleaching and Signal Stability in Fluorescence Imaging*

Fluorophores are vulnerable to photo bleaching under prolonged or intense irradiation. Minimize exposure duration and intensity while acquiring images. Employ antifade reagents within mounting media to extend fluorophore lifetime. Maintain samples at 4 °C and in dark conditions when not imaging. If available, use advanced illumination techniques such as light sheet microscopy or spinning disk confocal that reduce photo damage compared to epifluorescence.

(l) *Representative Controls and Validation*

Include unstained samples to assess background auto fluorescence inherent to biological materials or substrata. Use single-stain controls to calibrate spectral overlap and ensure fluorescence exclusivity among channels. Incorporate diffraction-limited microspheres with known dimensions and fluorescence properties for calibration and performance monitoring.

Competing Interest

The authors declare that there are no competing interests regarding the publication of this chapter.

References

1. Donlan RM (2002) Biofilms: microbial life on surfaces. Emerg Infect Dis 8(9):881–890
2. Neu TR, Lawrence JR (2015) Innovative techniques, sensors, and approaches for imaging biofilms at different spatial scales. FEMS Microbiol Ecol *91*(9):fiv113
3. O'Toole GA (2011) Microtiter dish biofilm formation assay. J Vis Exp 47:e2437
4. Bridier A, Briandet R, Thomas V, Dubois-Brissonnet F (2011) Resistance of bacterial biofilms to disinfectants: a review. Biofouling 27(9):1017–1032
5. Myszka DG, Südhof TC (2018) Fluorescence microscopy techniques for biofilm analysis. In: Methods in molecular biology, vol 1836, pp 25–38
6. Stewart PS, Franklin MJ (2008) Physiological heterogeneity in biofilms. Nat Rev Microbiol 6(3):199–210
7. Stoodley P, Sauer K, Davies DG, Costerton JW (2002) Biofilms as complex differentiated communities. Ann Rev Microbiol 56:187–209
8. Tolker-Nielsen T, Sternberg C (2011) Growing and analyzing biofilms in flow cells. Curr Protoc Microbiol 21(1):1B.2.1–1B.2.17
9. Hohmann-Marriott MF, Dubochet J (2014) Electron microscopy for the imaging of biofilms. Curr Opin Biotechnol 27:127–134
10. Neu TR (1996) Significance of bacterial surface-active compounds in interaction of bacteria with interfaces. Microbiol Mol Biol Rev 60(1):151–166
11. Heydorn A, Nielsen AT, Hentzer M, Sternberg C, Givskov M, Ersboll BK, Molin S (2000) Quantification of biofilm structures by the novel computer program COMSTAT. Microbiology 146(10):2395–2407
12. Torres-Sánchez A, Toftgaard Nielsen S (2018) Evaluation of LIVE/DEAD BacLight staining for flow cytometric differentiation of viable and non-viable cells. J Microbiol Methods 149:1–6
13. Abcam (n.d.) Crystal violet staining protocol. Retrieved June 10, 2025, from https://www.abcam.com/protocols/crystal-violet-staining-protocol
14. Sigma-Aldrich (n.d.) Safranin staining protocol. Retrieved December 31, 2024, from https://www.sigmaaldrich.com/technical-documents/protocols/biology/safranin-stain
15. O'Toole GA, Kolter R (1998) Initiation of biofilm formation in *Pseudomonas fluorescens* WCS365 proceeds via multiple, convergent signalling pathways: a genetic analysis. Mol Microbiol 28(3):449–461
16. Hung C, Guvener Z, Algebraic M et al (2007) A global regulator controls quorum sensing and biofilm formation in *Pseudomonas aeruginosa*. Mol Microbiol 65(1):107–121

Chapter 3

Fluorescence Microscopy Techniques for Biofilm Visualization and Quantification

Abirami J and Abirami K

Abstract

Biofilms are complex, surface-associated microbial communities embedded within a self-produced extracellular polymeric substance (EPS) matrix, playing critical roles in medical, environmental, and industrial settings. Fluorescence microscopy has emerged as a powerful and versatile tool for the visualization and quantitative analysis of biofilms due to its high sensitivity, specificity, and compatibility with a wide range of fluorescent probes. This chapter provides a comprehensive overview of fluorescence microscopy techniques employed for biofilm visualization and quantification, including wide-field fluorescence microscopy and confocal laser scanning microscopy (CLSM). Key fluorescent stains and probes used to assess biofilm biomass, spatial organization, viability, and matrix components are discussed. This chapter further outlines standardized methodologies for image acquisition, processing, and quantitative analysis, focusing on parameters such as fluorescence intensity, surface coverage, biofilm thickness, biovolume, and live/dead cell ratios. Emphasis is placed on maintaining consistency in imaging conditions and applying robust image analysis workflows using commonly available software tools. By integrating qualitative visualization with quantitative metrics, fluorescence microscopy enables detailed characterization of biofilm structure and dynamics, supporting its widespread application in biofilm research, antimicrobial evaluation, and surface-associated microbial studies.

Key words SYTO 9, Propidium iodide (PI), LIVE/DEAD BacLight assay, eDNA staining, Z-stack imaging

1 Introduction

Biofilms are structured communities of microorganisms enmeshed within a self-produced extracellular polymeric substance (EPS) that adhere to surfaces and interfaces. Their complex three-dimensional architecture and phenotypic heterogeneity make biofilms fundamentally different from planktonic cells, conferring enhanced tolerance to environmental stressors, antimicrobials, and host defense. Because of these properties, biofilms are implicated in chronic infections, industrial biofouling, and environmental processes,

Abirami K (ed.), *Microbial Biofilm Research: Comprehensive Protocols and Best Practices*, Springer Protocols Handbooks, https://doi.org/10.1007/978-1-0716-5535-1_3,

prompting intense research into effective detection and characterization strategies [1].

Fluorescence microscopy has emerged as a cornerstone technique for biofilm research due to its capacity to visualize living cells and matrix components with high specificity, probe structural features in situ, and enable quantitative analysis of biomass and spatial distribution. In contrast to conventional light microscopy, which primarily reveals gross morphology, fluorescence methods rely on fluorophores either fluorescent dyes or genetically encoded fluorescent proteins to label specific targets such as nucleic acids, proteins, polysaccharides, or indicators of cell viability. When excited at defined wavelengths, these fluorophores emit signals that can be detected and translated into high-contrast images of biofilm architecture, including live/dead cell distributions and EPS components. Among fluorescence techniques, confocal laser scanning microscopy (CLSM) stands out as especially powerful. By capturing optical sections at different depths through a biofilm and reconstructing them into three-dimensional images, CLSM allows researchers to visualize the spatial organization of cells and matrix, measure parameters such as biofilm thickness and roughness, and distinguish multiple components via multichannel fluorescence. Unlike electron microscopy, CLSM can be applied under hydrated, near-native conditions and in real time, preserving biological context while avoiding extensive sample preparation [2].

Modern fluorescence approaches extend beyond basic CLSM. For example, fluorescence in situ hybridization (FISH) and its advanced variants (e.g., CLASI-FISH, HiPR-FISH) combine sequence-specific probes with confocal imaging to map the spatial distribution of multiple microbial taxa within complex biofilms, revealing interspecies interactions that are otherwise obscured by bulk molecular methods. Additionally, wide-field epifluorescence microscopy coupled with image analysis can provide rapid, semi-quantitative assessments of biofilm surface coverage and relative biomass, offering practical utility in routine studies or comparative experiments with different conditions.

Beyond simple visualization, the fluorescence signal itself can be quantitatively analyzed using software tools that extract metrics such as biovolume, cell counts, viability ratios, and spatial heterogeneity. Recent developments in automated analysis platforms and biofilm cytometry tools have enhanced the reproducibility and throughput of such measurements, enabling statistically robust comparisons across experimental conditions. In summary, fluorescence microscopy has transformed how biofilms are detected and studied. Its combination of specificity, structural resolution, and adaptability to quantitative analysis makes it indispensable for contemporary biofilm research. The continual evolution of fluorescence techniques and analytical tools promises to further deepen our understanding of biofilm biology in both fundamental and applied settings [3].

2 Materials

2.1 Bacterial/Fungal Strains of Interest (Biofilm-Forming Organisms)

- Sterile culture media (e.g., LB broth, tryptic soy broth, and nutrient broth, depending on organism).
- Sterile 96-well plates, glass slides, or coverslips (surface for biofilm formation).
- Incubator (with shaking option if needed).

2.2 Fixation

- Paraformaldehyde or glutaraldehyde (to fix biofilm cells/structure).
- Phosphate-buffered saline (PBS) for washing.

2.3 Fluorescent Dyes/Probes (Depending on Study Focus)

2.3.1 General Cell Stains

- SYTO 9 (labels live bacteria, green).
- Propidium iodide (PI) (labels dead/damaged cells, red) → Often combined in the LIVE/DEAD Backlight kit.

2.3.2 Biofilm Extracellular Matrix (EPS) Stains

- Concanavalin A (ConA) conjugated with FITC, TRITC, or Alexa Fluor → binds polysaccharides.
- Calcofluor white → stains β-glucans (often used for fungal biofilms).
- Wheat Germ Agglutinin (WGA)-fluorophore conjugates → binds N-acetylglucosamine residues (e.g., in EPS).

2.3.3 Nucleic Acid Stains

- DAPI (blue, binds DNA).
- SYBR Green (DNA stain, green).

2.4 Microscopy and Detection

- Fluorescence microscope (epifluorescence or confocal laser scanning microscope, CLSM, preferred for 3D biofilm imaging).
- Appropriate filter sets/lasers for chosen dyes (e.g., DAPI = UV excitation, FITC = 488 nm, TRITC = 543–555 nm).
- Glass-bottom dishes or confocal-compatible slides (for high-resolution imaging).

2.5 Additional Reagents and Supplies

- Mounting medium (with or without antifade, e.g., ProLong Gold).
- Sterile PBS/buffer for washing steps.
- Sterile pipette tips, microtubes, forceps.
- Cover slips (for slide prep).
- Immersion oil (for microscopy objectives).

3 Methodology

3.1 Biofilm Visualization by FMS

1. Grow biofilms using 96-well microtiter plate static assays.
2. Fill wells of microtiter plates with 200 μL of fresh LB containing 1:10 dilution of overnight cultures, such that green and red fluorescent cells are mixed in a 1:10 ratio. This initial inoculum can be diluted as desired.
3. Alternatively, cells without fluorescence can be used as background, and red and green fluorescent cells could be added in lower amounts to visualize clonal cell arrangements. When including cells without fluorescent reporters, it is important to not use antibiotic while inoculating biofilms so that the cells without plasmids are not affected. As cells without plasmids may not have antibiotic resistance, antibiotics should not be used when the biofilm inoculum contains non-resistant cells.
4. Carefully break 18 × 18 mm cover glass into pieces big enough to fit inside the wells of a microtiter plate (*see* **Note 1**).
5. Broke cover glass into pieces by taking 3–4 cover glass squares at a time and using a glass slide as a solid support to break them into smaller, approximately rectangular pieces. Glass cutting tools can also be used if desired. Recommend sterilizing cover glass pieces by autoclaving. Recommend preparing 3–5 replicates (*see* **Notes** 2).
6. Alternatively, cover glass pieces can also be inserted in wider wells, e.g., 6-well plates. Wells should be filled such that the liquid covers at least three quarters of the height of these wells.
7. Insert cover glass pieces perpendicularly inside the wells for the biofilm to grow on. Use tweezer when handling cover glass pieces.
8. Incubate plates overnight (16–24 h) or for desired duration for biofilm formation without shaking at 37 °C.
9. Using a tweezer, carefully take out the cover glass piece from a well.
10. Lay this cover glass piece flat on a fresh glass slide. Label the glass slide with any necessary cell names or conditions that are being tested.
11. Cover the glass piece with formed biofilm using a fresh 22 × 50 mm microscope cover glass.
12. Fix the 22 × 50 mm cover glass on the glass slide by taping its edges.
13. Set up glass slide, prepared in steps 6–9, on the microscope for imaging.

14. After focusing, capture a wide-field view of the biofilms using red and green fluorescence channels. Image biofilms at high magnification to resolve their structures at the cellular level (*see* **Notes 3**).
15. Using a tweezer, take a piece of cover glass with formed biofilm out of the well of microtiter plate.
16. Hold the cover glass perpendicularly over an agarose pad[2] made of phosphate-buffered saline (PBS).
17. Wash the biofilm by pipetting ~50 μL (or desirable volume depending on size of substrate) of PBS over the cover glass. Let the liquid run-off from washing fall on the agarose pad [4].
18. Cover the agarose pad with a fresh piece of microscope cover glass. Image using DIC and fluorescence (red and green channels) imaging modalities to analyze the washed biofilm (Fig. 1).

3.2 Biofilm Quantifications

3.2.1 Fluorescence Intensity

1. Fluorescence images are captured using identical exposure time, gain, laser power, and binning settings, and at least 5–10 random fields were acquired per sample.
2. The images were converted to 8-bit format (Image → Type → 8-bit), followed by background subtraction using the rolling ball algorithm with a radius of 20–50 pixels (Process → Subtract Background).
3. A region of interest (ROI) encompassing the biofilm area is defined, and fluorescence intensity is quantified using the *Analyze → Measure* function in Image.
4. The mean gray value was recorded as arbitrary units (AU) and normalized either to the analyzed area ($AU/\mu m^2$) or to the corresponding control sample to obtain fold-change values (*see* **Note 3**).
5. The results were expressed as mean fluorescence intensity (AU ± SD), representing relative biofilm biomass rather than absolute cell numbers.

3.2.2 Biofilm Coverage (%)

1. Biofilm surface coverage was quantified as a measure of surface colonization efficiency.
2. Fluorescence images were converted to grayscale and subjected to background subtraction prior to analysis.
3. Thresholding was performed using the *Image → Adjust → Threshold* function in ImageJ, and identical threshold settings were applied to all samples so that the biofilm regions were consistently highlighted.
4. The area occupied by the biofilm was measured after selecting *Analyze → Set Measurements* and enabling the area parameter, followed by *Analyze → Measure*.

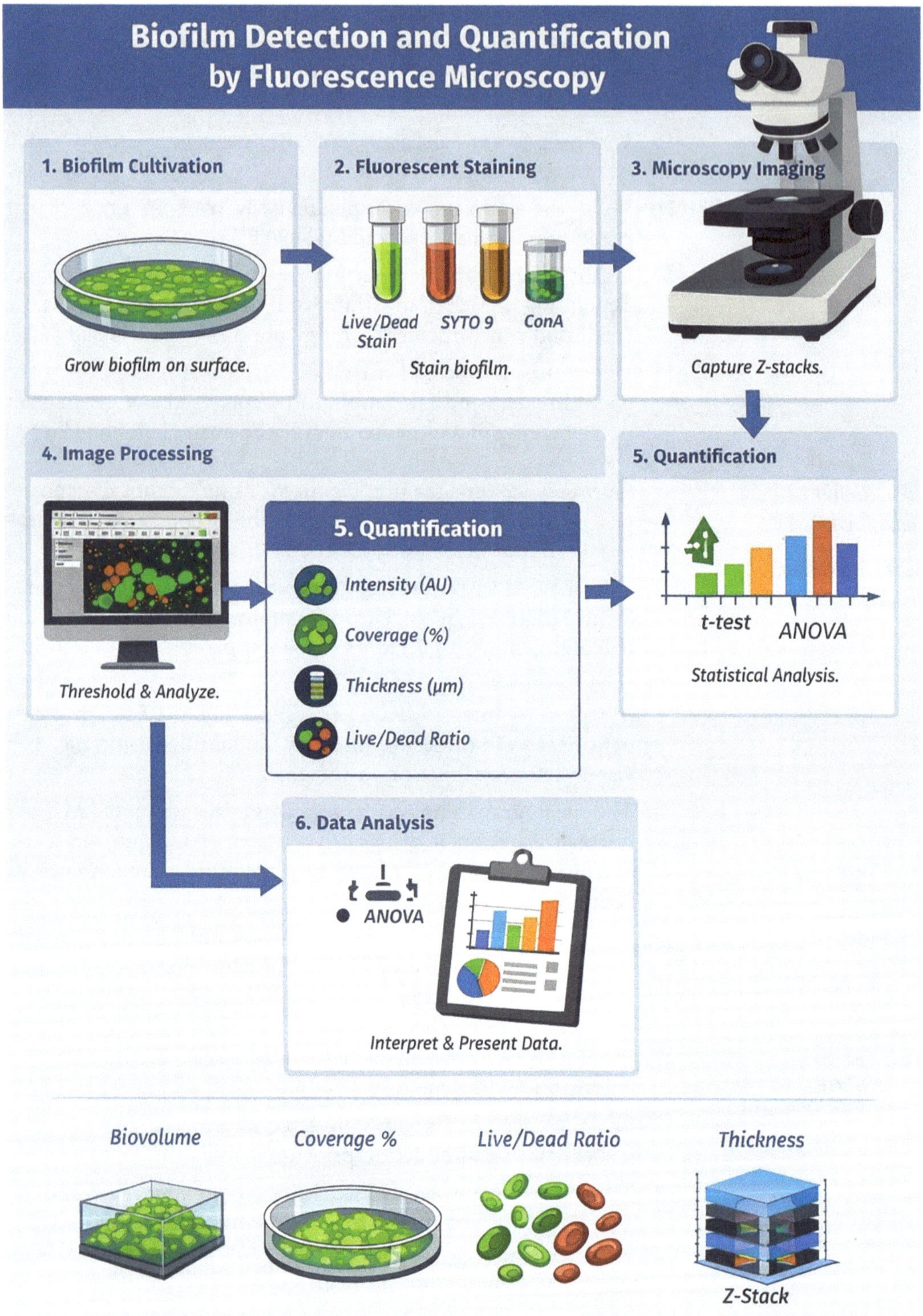

Fig. 1 Stepwise protocol for biofilm detection and quantification

5. The total image area was also measured, and biofilm coverage was calculated as the ratio of biofilm area to total area multiplied by 100. The results were expressed as percent surface coverage (% ± SD) (*see* **Notes 4**).

   Calculation

$$\text{Biofilm coverage } (\%) = \frac{\text{Biofilm area}}{\text{Total area}} \times 100$$

   Output

   Percent surface coverage (% ± SD)

3.2.3 Biofilm Thickness (m)

1. Biofilm thickness is determined using confocal laser scanning microscopy (CLSM) Z-stack analysis.
2. Optical sections were acquired at fixed Z-intervals (e.g., 0.5–1 μm), beginning at the substratum and extending above the biofilm.
3. The Z-stack images are imported into ImageJ, and Z-projection was performed when required for visualization (*Image → Stacks → Z-Project*).
4. Biofilm thickness was calculated by identifying the first and last optical sections showing detectable fluorescence and multiplying the total number of slices by the Z-step size (*see* **Notes** 5).

$$\text{Thickness } (\mu\text{m}) = \text{Number of slices} \times \text{Z} - \text{step}$$

   Alternative: Use COMSTAT for mean and maximum thickness.

   Output: Mean biofilm thickness (μm ± SD).

3.2.4 Biovolume (m^3/m^2)

1. Three-dimensional biomass is quantified by calculating biovolume, which represents the total biofilm biomass volume normalized to the surface area and serves as a robust 3D metric.
2. For this analysis, fluorescence Z-stack images were thresholder slice by slice, and the image stack was converted into binary images.
3. The biofilm area in each optical section was measured, multiplied by the corresponding Z-step thickness to obtain the volume for each slice, and the volumes from all slices were summed to determine the total biovolume (*see* **Notes** 6).

$$\text{Biovolume} = \frac{\sum(\text{Area}_Z \times \text{Z} - \text{step})}{\text{Substratum area}}$$

4. Software Options.
   (a) COMSTAT/COMSTAT2.
   (b) BiofilmQ.
   (c) Imaris.
5. Output: Biovolume ($\mu m^3/\mu m^2$).

3.2.5 Live/Dead Ratio (Cell Viability Assessment)

1. Cell viability within the biofilm was assessed by determining the live/dead ratio using dual-fluorescence staining, in which live cells were stained green with SYTO 9 and dead cells were stained red with propidium iodide.
2. Fluorescence images were acquired separately for the green and red channels using appropriate filter sets to avoid spectral overlap.
3. The acquired images were processed in ImageJ by splitting the channels (*Image → Color → Split Channels*), followed by independent background subtraction for each channel.
4. Identical threshold settings were applied to both channels, and the mean fluorescence intensity of each channel was measured; alternatively, the area occupied by live and dead cell populations was quantified to calculate the live/dead ratio (*see* **Notes** 7).

   Calculation

$$\text{Live/Dead Ratio} = \frac{\text{Green fluorescence (AU)}}{\text{Red fluorescence (AU)}}$$

   Output

   Ratio or percentage viability:

$$\%\text{Live cells} = \frac{\text{Green}}{\text{Green} + \text{Red}} \times 100$$

4 Notes

1. Weak/No Fluorescence

   Causes: Low biomass, low dye concentration, photobleaching, wrong filter, plasmid loss (GFP).

   Fix: Increase growth time, titrate dye, reduce light exposure, verify filters, include positive control.

2. High Background Signal

   Causes: Excess dye, incomplete washing, media autofluorescence, plastic fluorescence.

   Fix: Wash 2–3× with PBS, use phenol red–free media, include unstained control, switch to glass if needed.

3. Uneven/Patchy Staining

   Causes: Thick biofilm, poor dye penetration, detachment during washing.

   Fix: Shorter growth time, gentle handling, use confocal Z-stacks.

4. LIVE/DEAD Interpretation Issues
   - Green ≠ always viable.
   - Red may indicate stress, not complete death.
   - Channel overlaps possible.

   Fix: Include heat-killed control, quantify red/green ratio, confirm with CFU counts.
5. Photobleaching

   Fix: Lower laser intensity, minimize repeated imaging, use antifade reagent and avoid long exposure times.
6. Autofluorescence

   Sources: Media components, metabolites, plastic surfaces.

   Fix: Image unstained control, compare emission spectra, consider far-red dyes.
7. Key Best Practices
   - Always include positive and negative controls.
   - Optimize dye concentration (don't overload).
   - Use consistent imaging settings.
   - Minimize biofilm disturbance.
   - Validate fluorescence data with complementary methods (e.g., CFU).

Competing Interest

The authors declare that there are no competing interests regarding the publication of this chapter.

References

1. Beyenal H, Donovan C, Lewandowski Z, Harkin G (2004) Three-dimensional biofilm structure quantification. J Microbiol Methods 59(3): 395–413, ISSN 0167-7012. https://doi.org/10.1016/j.mimet.2004.08.003
2. Frühauf HM, Stöckl M, Holtmann D (2022) R-based method for quantitative analysis of biofilm thickness by using confocal laser scanning microscopy. Eng Life Sci 22(6):464–470. https://doi.org/10.1002/elsc.202200008
3. Shailaja A, Bruce TF, Gerard P, Powell RR, Pettigrew CA, Kerrigan JL (2022) Comparison of cell viability assessment and visualization of *Aspergillus niger* biofilm with two fluorescent probe staining methods. Biofilms 4:100090. https://doi.org/10.1016/j.biofilm.2022.100090
4. Mountcastle SE, Vyas N, Villapun VM et al (2021) Biofilm viability checker: an open-source tool for automated biofilm viability analysis from confocal microscopy images. NPJ Biofilms Microbiomes 7:44. https://doi.org/10.1038/s41522-021-00214-7

Chapter 4

Microscopy-Based Methods: Visualizing Biofilm Architecture and Structure with a Focus on Confocal Laser Scanning Microscopy (CLSM)

Srinivasan P, Rajalakshmi M, and Monika Y

Abstract

Biofilms are microbial communities that adhere to surfaces and are embedded within a matrix of extracellular polymeric substances (EPS) that they synthesize themselves, which greatly enhances their ability to withstand environmental challenges and antimicrobial agents. Understanding how these communities are organized in three dimensions—such as how cells aggregate, how fluid channels are arranged, how matrix components are distributed, and how the biomass is stratified along the vertical axis—is essential for interpreting biofilm behavior in clinical, environmental, industrial, and food-related contexts. Confocal laser scanning microscopy (CLSM) has become a preferred imaging technique for biofilms because it permits nondestructive, three-dimensional visualization of intact, hydrated communities with micron-scale resolution by optically sectioning throughout the biofilm depth. This chapter presents a detailed, protocol-focused overview of how to visualize biofilm architecture with CLSM, primarily for bacterial systems but with straightforward adaptations for fungal and mixed-species biofilms. This chapter addresses conceptual principles and specific advantages of CLSM in biofilm research; comparison and selection of growth formats such as static systems and flow cells; rationale for selecting and handling fluorescent stains and probes for cells and EPS; and step-wise procedures for biofilm cultivation, labeling, mounting, and imaging. Recommendations are provided for configuring CLSM acquisition parameters—including choices of excitation and emission channels, pinhole settings, Z-step intervals, and time-lapse options to obtain high-quality, reproducible datasets. Observation frameworks are outlined to help readers systematically evaluate (i) global morphology and biomass distribution, (ii) spatial patterns of viable and nonviable cells, (iii) localization of matrix constituents, and (iv) temporal changes in structure during biofilm development or after treatment. This chapter concludes with a brief guide to integrating CLSM datasets with quantitative image analysis methods to obtain architectural metrics such as biovolume, overall and maximum thickness, roughness, and surface coverage, thereby linking microscopic observations to biologically meaningful interpretations.

Key words Biofilm architecture, Extracellular polymeric substances (EPS), Confocal laser scanning microscopy (CLSM), Three-dimensional biofilm imaging, Fluorescent staining of biofilms, Flow-cell biofilm model, Live/dead viability assays, EPS matrix visualization, COMSTAT image analysis, Quantitative biofilm morphology

Abirami K (ed.), *Microbial Biofilm Research: Comprehensive Protocols and Best Practices*, Springer Protocols Handbooks, https://doi.org/10.1007/978-1-0716-5535-1_4,

1 Introduction

1.1 Biofilm Architecture and Its Significance

Biofilms are organized assemblies of microorganisms that colonize a surface and are embedded in an EPS matrix composed predominantly of polysaccharides, proteins, lipids, and extracellular DNA. The spatial arrangement of this assemblage—including its thickness, the fraction of the surface it occupies, the size and shape of microcolonies, and the presence of internal channels—directly shapes mass transport, signaling, and stress tolerance within the community. Gradients in oxygen, nutrients, and waste products frequently arise across the depth of the biofilm, giving rise to heterogeneous microenvironments that support physiologically distinct subpopulations; these stratified niches are key contributors to biofilm persistence in medical devices, chronic infections, water systems, and industrial pipelines.

Conventional bulk assays, such as CFU enumeration, crystal violet staining of attached biomass, or metabolic activity indicators, collapse this heterogeneity into a single readout and cannot resolve how cells and matrix are organized within the structure. Microscopic approaches, in contrast, provide spatially resolved information about how cells distribute, cluster, or disperse across and within the matrix. Bright-field and wide-field fluorescence microscopy can reveal basic patterns of attachment and clustering but suffer from out-of-focus blur and limited optical sectioning capacity, especially when imaging thick, hydrated biofilms. Electron microscopic techniques produce high-resolution, ultrastructural information but require dehydration, fixation, and other sample preparation steps that can substantially alter the native configuration of the hydrated matrix and channels [1].

1.2 Confocal Laser Scanning Microscopy for Biofilms

Confocal laser scanning microscopy operates by illuminating only a small point or line within the sample and using a spatial pinhole in front of the detector to exclude light originating from planes above and below the focal plane. By sequentially acquiring optical sections along the Z-axis and digitally stacking these images, CLSM yields three-dimensional datasets (Z-stacks) that can be re-oriented, rendered volumetrically, and used for quantitative morphometric analysis. When biofilms are labeled with suitable fluorescent probes—either genetically encoded fluorophores or chemical stains directed to cells and matrix components—CLSM supports simultaneous, multichannel imaging of live and dead cells, EPS polymers, and extracellular DNA within the intact community.

Multiple experimental studies have used CLSM to map biofilm architecture and matrix composition in canonical bacterial systems such as *Pseudomonas aeruginosa* and *Staphylococcus aureus*, as well as in environmental and clinical isolates. Time-resolved CLSM imaging in flow cells and microfluidic devices allows researchers to follow the entire biofilm life cycle, including initial adhesion, growth and maturation, channel formation, and dispersal, and to observe how exposure to antimicrobials or changes in hydrodynamics affect structural organization. In parallel, a range of freely available and commercial image analysis tools—such as COMSTAT, FIJI/ImageJ plugins, and advanced 3D analysis platforms—enable segmentation of CLSM Z-stacks to extract quantitative descriptors, including total biovolume, average and maximal thickness, surface coverage, and roughness coefficients [2].

The central objective of this chapter is to provide a cohesive set of experimental and interpretive guidelines for using CLSM to visualize and analyze biofilm architecture. The target audience includes microbiologists, technologists, and students who plan to implement confocal imaging in research or teaching laboratories. Emphasis is placed on clear, stepwise protocols and decision points—from culture preparation and staining through acquisition and structural interpretation—so that the methods can be adapted to a broad range of organisms, surfaces, and resource settings [3].

2 Materials Required

2.1 Biofilm Growth Systems and General Microbiology Supplies

The choice of cultivation system should be driven by the biological question, the level of environmental control required, and the available instrumentation. Commonly used configurations include the following.

2.1.1 Static Microtiter-Based Systems

- Sterile, flat-bottom 24-well or 96-well polystyrene plates for parallel cultivation.
- Sterile glass coverslips (No. 1.5) or chambered cover glass inserts placed in wells when glass surfaces or direct imaging in the growth chamber are desired.

2.1.2 Flow-Based Systems

- Commercial flow-cell chambers (e.g., three-channel flow cells) with glass or optically clear plastic bottoms, or custom-fabricated chambers in which a coverslip serves as the imaging window.
- A peristaltic pump and appropriate tubing (e.g., silicone) to deliver medium at controlled flow rates.
- Sterile medium reservoirs and waste containers integrated into a closed-loop system.

2.1.3 Coupon or Reactor-Based Systems

- Coupons fabricated from glass, stainless steel, or application-specific materials (such as catheter segments or food-processing surfaces) mounted within flow reactors or standardized CDC biofilm reactors.

2.2 Standard Microbiological Materials

- Bacterial strains or other microorganisms of interest (e.g., *P. aeruginosa*, *S. aureus*, or environmental communities), with appropriate biosafety measures.
- Culture media formulated for the organism and application, such as LB or TSB for general growth, defined minimal media, or simulation media (e.g., artificial saliva or urine for host-associated models).
- Sterile phosphate-buffered saline (PBS) or physiological saline (0.9% NaCl) solutions for rinsing and dilutions.
- Sterile culture flasks, tubes, micropipette tips, and serological pipettes for routine handling.

2.3 Confocal Microscope and Related Imaging Hardware

- An inverted or upright CLSM platform equipped with:
  - Laser lines suited to the excitation spectra of planned fluorophores (commonly 405, 458, 488, 514, 561, and 633 nm).
  - High-numerical-aperture objectives such as 40× water immersion, 63× oil immersion, and 20× or 40× air objectives for survey imaging.
  - A motorized stage and Z-drive with micrometer or submicrometer resolution for precise control of Z-step size.
  - Optional environmental controls, including a heated stage insert and, when required, CO_2 and humidity regulation for prolonged live imaging.
- Proprietary imaging software supplied with the microscope (e.g., Zeiss ZEN, Leica LAS X, Olympus FV, Nikon NIS-Elements), capable of multichannel acquisition, Z-stack collection, and time-lapse imaging.

2.4 Mounting Accessories

- Standard glass slides and coverslips (No. 1.5) matched to the objective specifications and immersion medium.
- Chambered cover glass systems that allow direct growth and imaging of the biofilm in the same vessel.
- Silicone spacers or readymade imaging chambers designed to preserve the thickness and integrity of thick biofilms.
- Nail polish or biocompatible sealants to close chamber edges for long imaging sessions or to reduce evaporation during live imaging.

2.5 Fluorescent Stains and Probes

Fluorophore selection must reflect the architectural features and components being investigated, while also considering spectral compatibility in multichannel experiments.

(a) *Viability (Live/Dead) Indicators*

- Commercial "Live/Dead" bacterial viability kits based on SYTO 9 (or similar green nucleic acid stain) combined with propidium iodide (PI), which penetrates cells with compromised membranes.

- Alternative nucleic acid dyes from the SYTO family or related series that can be paired with PI or other membrane-impermeant probes to distinguish intact from damaged cells.

(b) *General Nucleic Acid Stains (Total Cells and Extracellular DNA)*

- SYTO dyes, SYBR Green, TOTO/TOTO-3, and comparable high-affinity DNA-binding fluorochromes for global visualization of cellular nucleic acids.
- DAPI or Hoechst dyes, typically excited at 405 nm, that can label chromosomal DNA and, under some staining conditions, extracellular DNA within the matrix.

(c) *EPS and Matrix-Targeted Probes*

- Fluorescent lectin conjugates that bind specific carbohydrate motifs (e.g., wheat germ agglutinin for N-acetylglucosamine-rich polymers; concanavalin A for selected glucose- and mannose-containing structures).
- Labeled dextrans that function as inert tracers to outline water channels and assess pore connectivity or matrix porosity.
- Dyes such as thioflavin-T or related probes for identifying amyloid-like protein fibrils that contribute to the structural scaffold of certain biofilms.

(d) *Reporter Strains*

- Bacterial strains engineered to constitutively or inducibly express fluorescent proteins (GFP, YFP, mCherry, DsRed, and others), which allow visualization of specific subpopulations, gene expression patterns, or strain mixtures without external staining.

(e) *Stain Preparation and Storage*

- Prepare concentrated stock solutions in sterile water, PBS, or DMSO as specified in manufacturer or literature protocols.
- Store aliquots at recommended temperatures and protect from light exposure to minimize degradation and photo bleaching, avoiding repeated freeze-thaw cycles.

2.6 Additional Reagents and Consumables

- Sterile distilled or deionized water for solution preparation.
- Fixatives for endpoint imaging, such as 2–4% paraformaldehyde (PFA) in PBS, and low-percentage glutaraldehyde when required for specific structural preservation.

- Optional optical-clearing reagents or high-refractive-index aqueous media to improve imaging depth and contrast in very thick or dense biofilms.
- Parafilm, aluminum foil for light protection, and micro centrifuge tubes for handling staining solutions and fixed samples.

2.7 Software for Visualization and Image Analysis

- FIJI/Image J, supplemented with 3D visualization, segmentation, and analysis plugins, for initial inspection and basic quantification.
- COMSTAT or COMSTAT2 implementations (often as scripts or plug-ins) tailored for extracting standard biofilm structural metrics from CLSM Z-stacks.
- Manufacturer-provided or third-party 3D rendering and analysis tools (e.g., ZEN 3D modules, Imaris, Amira) for advanced visualization and quantitative workflows.

2.8 General Considerations Before Starting

2.8.1 Experimental Design

- Clearly define the central question before setting up the experiment, such as comparing architectures between strains, evaluating the structural impact of antibiotics, or assessing how surface chemistry alters matrix distribution.
- Match the culture model to the question: static plate-based formats are convenient and high-throughput, whereas flow cells more realistically reproduce shear forces and continuous nutrient supply.
- Decide early whether the main emphasis will be on live imaging (including time-lapse sequences) or endpoint analysis; this decision influences stain choice, sample handling, and whether fixation is appropriate.

2.8.2 Controls

- Always prepare unstained biofilms to assess background autofluorescence from cells, medium components, and substratum materials.
- Use thin smears or early-stage biofilms as test samples for optimizing fluorescent dye concentrations, incubation times, and acquisition parameters before imaging complex, mature biofilms.
- For viability assays, incorporate well-defined live and dead controls—such as untreated cultures and ethanol- or heat-killed cells—to set channel ranges, establish thresholds, and validate interpretation of live/dead ratios.

3 Methods

3.1 Protocol 1: Preparation of Static Biofilms on Glass Coverslips for CLSM

This protocol outlines an accessible approach for generating biofilms on coverslips inserted into multiwell plates, followed by staining and CLSM imaging. The method is readily adaptable to teaching laboratories and routine research applications.

Step 1: Preparation of the Inoculum

1. Streak the microbial strain onto an appropriate agar medium and incubate until well-isolated colonies appear.
2. Transfer a single colony into 5–10 mL of liquid medium and incubate with agitation until the culture reaches mid-exponential phase (e.g., OD_{600} around 0.4–0.6).
3. Dilute this culture into fresh medium so that the starting cell density is standardized, such as $OD_{600} \approx 0.05$ or approximately 10^7 CFU/mL, adjusting as needed for species-specific growth and the desired biomass at the imaging time point (*see* **Note 1**).

Step 2: Preparation of Coverslips and Wells

1. Clean coverslips using ethanol or suitable detergents, rinse thoroughly with distilled water, and allow to dry; sterilize by autoclaving or flaming as appropriate.
2. Using sterile forceps, place one coverslip at the bottom of each selected well of a 24-well or 6-well plate.
3. Add the standardized inoculum to each well (e.g., 1–2 mL per well in 24-well plates), ensuring that the entire coverslip surface is submerged.
4. Include control wells containing medium alone to monitor sterility and background signals.

Step 3: Biofilm Cultivation

1. Incubate the plate under the appropriate temperature con ditions (commonly 30–37 °C) without shaking for 24–72 h, depending on the organism and the stage of biofilm development of interest.
2. For extended growth periods, gently replace a fraction of the spent medium at defined intervals with fresh medium, taking care not to disturb the attached biofilm on the coverslip.
3. If a time-course experiment is planned, set up replicate wells for each target time point so that each coverslip is only handled once (*see* **Note 2**).

Step 4: Washing and Optional Fixation

1. Remove the planktonic culture carefully from each well by aspirating liquid from the side of the well, avoiding direct contact with the coverslip surface.
2. Add sterile PBS or saline to each well and gently aspirate and replace this wash solution two to three times to remove nonadherent cells and residual medium.
3. For live-cell imaging, proceed directly to staining after washing.
4. For fixed preparations, carefully add enough 2–4% PFA in PBS to fully cover the coverslip, incubate for 15–20 min at room temperature, and then wash two or three times with PBS to eliminate residual fixative (*see* **Note 3**).

3.2 Protocol 2: Staining Strategies for Biofilm Visualization

A. *Live/Dead Viability Staining*

Viability staining provides a rapid visual overview of how viable and nonviable cells are distributed throughout the biofilm and is frequently used to assess the impact of antimicrobial exposure or environmental stress.

1. Prepare the working dye solution according to the viability kit instructions, typically mixing SYTO 9 (or equivalent green-fluorescent nucleic acid dye) with PI in PBS at the recommended concentrations.
2. Add enough staining solution to each well to immerse the coverslip (e.g., 0.5–1 mL in a 24-well plate).
3. Incubate the plate in the dark at room temperature for 15–20 min to allow the dyes to penetrate the biofilm structure.
4. Gently rinse once with PBS to remove excess unbound dye, taking care not to shear off the biofilm.
5. Leave the stained biofilm in a small volume of PBS or appropriate imaging buffer for confocal examination.

   Typical detection channels:

- Green fluorescence channel for cells with intact membranes labeled by SYTO 9 or similar dyes, usually excited at ~488 nm and collected in a ~500–550 nm emission window (see **Note 4**).
- Red fluorescence channel for cells with compromised membranes stained by PI, typically excited near 561–568 nm and detected in the ~580–650 nm emission range.

B. *Total Biomass and Nucleic Acid Staining*

When discrimination between live and dead cells is not essential, or when imaging fixed samples, single nucleic acid stains offer a simpler approach to visualize overall cellular biomass.

1. Dilute the chosen nucleic acid dye (e.g., SYTO series, SYBR Green, or DAPI) to its working concentration in PBS.
2. Incubate the washed or fixed biofilms with this solution in the dark for 10–20 min.
3. Briefly wash with PBS and maintain the biofilms in buffer for imaging (*see* **Note 5**).

These labels can be combined with matrix-specific stains in other spectral channels to obtain multicomponent images of cells and EPS.

C. *Matrix Polysaccharide and Extracellular Component Staining*

To map the distribution of EPS:

1. Select lectin-fluorophore conjugates that bind known or suspected polysaccharide components of the matrix. For instance, wheat germ agglutinin is useful for detecting N-acetylglucosamine-containing polymers, whereas concanavalin A targets selected mannose- and glucose-rich structures.
2. Prepare the lectin solutions in the appropriate buffer at the manufacturer's recommended concentrations, including cofactors such as calcium ions if required for binding.
3. Incubate the biofilms with the lectin solution for 20–30 min in the dark, ensuring that the entire specimen is covered.
4. Gently wash two to three times with buffer to remove unbound lectin and reduce background signal.
5. Optionally, combine lectin staining with a distinct nucleic acid or viability marker to visualize cells and matrix simultaneously in separate channels.

For extracellular DNA:

- Use DNA-binding dyes under conditions that allow labeling of both cellular and extracellular DNA, and interpret staining patterns relative to cell positions and void regions; where possible, complementary strategies can be used to differentiate intracellular from extracellular DNA pools.

D. *Reporter Strains and Multichannel Imaging*

Genetically encoded reporters provide powerful tools for tracking specific populations or gene expression states without the need for exogenous dyes. When these strains are combined with additional fluorescent stains, attention must be paid to spectral separation, filter selection, and the potential need for sequential scanning or spectral unmixing to minimize channel cross-talk.

3.3 Protocol 3: Mounting of Coverslips and Sample Handling for CLSM

1. After staining, carefully retrieve each coverslip from the well using blunt, sterile forceps, gripping only the edges to avoid scraping the biofilm.
2. Place a small droplet of imaging buffer (PBS or another suitable medium) on a clean glass slide or inside a pre-formed silicone spacer affixed to the slide.
3. Invert the coverslip so that the biofilm faces downward onto the droplet, lowering it slowly to minimize air bubble formation.
4. If a spacer is used, gently press the coverslip to ensure good contact between the buffer and the biofilm while avoiding compression of thick structures.
5. Seal the margins of the coverslip with nail polish or a compatible sealant, if long imaging sessions or transport to another facility is anticipated.
6. Protect the prepared slide from light and evaporative drying, and move it to the microscope stage as soon as practical to maintain fluorescence intensity and structural integrity.

For live time-lapse experiments, it is often preferable to grow and image biofilms directly in chambered cover glass systems or perfusion chambers that can be mounted on a temperature-controlled stage, while medium and gases are continuously supplied [4].

3.4 Protocol 4: CLSM Acquisition Settings for Biofilm Architecture

A. *Microscope Setup and Focusing*

1. Select an objective lens appropriate for the sample thickness and refractive index: water immersion objectives (e.g., 40×) are advantageous for thick, aqueous biofilms, whereas high-NA oil immersion objectives (e.g., 63×) provide superior resolution for near-surface structures.
2. Apply the correct immersion medium and confirm that the software settings for coverslip thickness and refractive index match the actual sample conditions.
3. Use transmitted light or low-intensity fluorescence excitation to identify the coverslip plane and locate representative regions of the biofilm.
4. Adjust focus and detector gain to obtain sharp images with adequate signal while avoiding overexposure and detector saturation that would compromise quantitative analysis.

B. *Selection of Imaging Parameters*

Laser Lines and Detectors

- Choose excitation wavelengths and emission filter ranges that match the absorption and emission characteristics of each fluorophore, minimizing spectral overlap among channels.

- Where overlaps cannot be fully avoided, employ sequential scanning or advanced spectral unmixing algorithms, if available, to reduce bleed-through.

Pinhole Size

- Set the confocal pinhole to approximately one Airy unit per channel to achieve a suitable compromise between axial resolution and signal intensity.

Scan Format and Averaging

- Select image dimensions (e.g., 512 × 512 or 1024 × 1024 pixels) to meet Nyquist sampling requirements for the chosen objective and desired resolution.
- Use line or frame averaging when necessary to improve signal-to-noise ratios, being mindful that higher averaging increases acquisition time and contributes to photobleaching.

Z-Stack Configuration

- Determine the bottom of the biofilm as the first plane showing clear fluorescence above the glass surface and the top as the last plane with detectable biomass.
- Choose a Z-step size at or below the axial resolution limit (often 0.3–1.0μm for high-NA lenses) to capture vertical structure accurately without unnecessary oversampling.
- Include some buffer region above and below the biofilm to define boundaries and facilitate later analysis (*see* **Note 6**).

Time-Lapse (4D) Imaging

- Set time intervals that balance the need for temporal resolution with concerns about phototoxicity, bleaching, and data volume.
- Minimize laser power, exposure, and scanned area when capturing repeated images of the same location to reduce stress on the biofilm and preserve physiological relevance [Fig. 1].

C. *Acquisition Workflow*

1. Begin with low-magnification overview scans to identify areas that represent the typical morphology of the sample, avoiding regions at the extreme edges or with obvious artifacts.
2. Acquire Z-stacks at predetermined positions within each condition, ensuring that imaging parameters are held constant between replicates and treatments to enable valid comparisons.

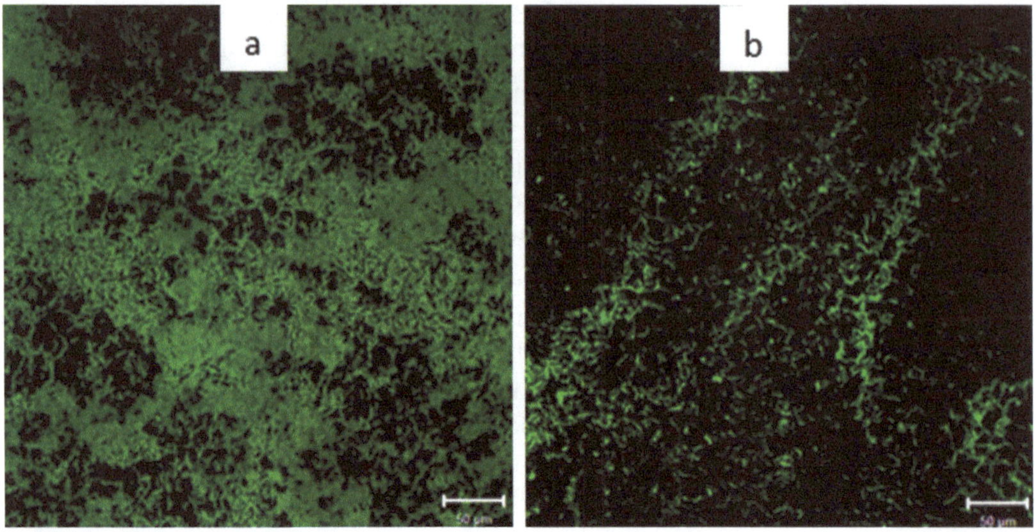

Fig. 1 Confocal laser scanning microscope 2D images (×200) of *E. coli* biofilm. (**a**) Control biofilm and (**b**) biofilm treated with selected antibiotics at BIC, respectively. Untreated test pathogens served as control biofilm. Images were taken using CLSM microscope (Scale bar = 50μm) LSM-170, Carl Zeiss Jena, Germany

3. Save raw datasets in high-bit-depth formats with full metadata, avoiding lossy compression so that quantitative analysis remains reliable.
4. Record detailed information on stain concentrations, incubation times, objective and laser settings, and Z-stack parameters for documentation and reproducibility.

3.5 Observation Strategies for Biofilm Architecture

A structured approach to interpreting CLSM datasets helps convert complex 3D images into biologically meaningful insights.

3.5.1 Overall Morphology and Biomass Distribution

Using 3D renderings, maximum-intensity projections, and orthogonal slice views:

- Evaluate *surface coverage* by examining how much of the substratum is occupied by biomass, distinguishing between isolated microcolonies, patchy distribution, and nearly continuous films.
- Estimate *biofilm thickness* from the distance between the bottom and top of the fluorescent signal, and qualitatively classify the architecture as layer-like, mushroom-shaped, or composed of tall aggregates separated by low-density regions (*see* **Note** 7).
- Assess *structural roughness* by observing variations in height across the field: highly heterogeneous structures will show pronounced peaks and troughs, whereas smoother biofilms display more uniform topography.

Relating these patterns to growth conditions can reveal how nutrient level, shear, and surface properties influence structural development—for instance, strong shear and high nutrients may promote pronounced tower and streamer formation, whereas nutrient limitation often favors thinner, more uniform layers.

3.5.2 Spatial Patterns of Cell Viability (Live/Dead Architecture)

In biofilms stained with live/dead reagents:

- Inspect where red-stained, membrane-compromised cells accumulate: near the surface exposed to the bulk fluid, in the central core, or at the interface with the substratum.
- Look for *vertical gradients* in viability, which commonly manifest as metabolically active cells in regions with greater nutrient and oxygen availability and more nonviable or dormant cells in deeper or more shielded zones.
- After exposure to antimicrobials, compare how different agents affect spatial killing: some compounds may primarily damage cells at the biofilm periphery, whereas others can penetrate and affect deeper layers (*see* **Note 8**).

XZ and YZ orthogonal reconstructions are especially informative for visualizing vertical viability patterns, and combining these with segmentation-based quantification allows depth-resolved live/dead ratios to be calculated.

3.5.3 Matrix Localization and Architecture

Matrix-specific probes, such as lectin conjugates, reveal how EPS components are arranged relative to the cellular framework and to each other.

Key questions include:

- Does the labeled matrix form a diffuse halo around single cells, concentrate tightly around microcolonies, or form a continuous scaffold that interconnects separate clusters?
- Are particular polysaccharides enriched at specific interfaces, such as near the substratum (supporting adhesion) or at the biofilm-fluid boundary (potentially contributing to protection or interactions with the environment)?
- Do certain matrix signals co-localize with distinct subpopulations, such as specific strains in a co-culture or cells in defined physiological states, indicating specialized micro-niches?

Comparing wild-type strains with mutants deficient in particular matrix components can link observed changes in biovolume, thickness, architecture (e.g., loss of mushroom-like structures), and detachment behavior to the absence of those matrix determinants (*see* **Note 9**).

3.6 Channel Networks and Void Spaces

Bright-field, inverted-intensity fluorescence images, and inert tracer studies together help identify and characterize water channels and interstitial voids.

Important aspects to examine:

- The presence and spatial distribution of channels—whether they extend from the substratum to the outer surface or are confined to certain depth intervals within the biofilm.
- The alignment of channels with the direction of flow in flow-cell systems, which may reflect the influence of hydrodynamics on structural organization and mass transport.
- Temporal changes in channel width, continuity, or connectivity after environmental perturbations or therapeutic interventions.

Channel networks play central roles in delivering nutrients and removing waste, as well as in mediating dispersal and penetration of antimicrobial compounds, so their visualization and, where possible, quantitative characterization provide essential context for functional studies.

3.7 Temporal Dynamics and Developmental Stages

Time-resolved CLSM imaging in flow-based or microfluidic systems allows direct observation of biofilm development.

- Monitor transitions from single-cell attachment to microcolony formation, early aggregates, and mature three-dimensional structures, while correlating these stages with changes in matrix deposition and surface coverage.
- Observe dispersal phenomena, such as the release of aggregates, erosion of outer layers, or sudden expansion of channels, and relate these events to environmental cues, nutrient shifts, or quorum-sensing signals.
- Compare regulatory mutants (e.g., quorum-sensing-defective or altered cyclic-di-GMP signaling) with parental strains to determine how signaling networks modulate architectural traits, including aggregation, matrix production, and detachment.

Synchronizing imaging schedules with controlled environmental manipulations (e.g., adding antibiotics or altering flow regimes at defined times) strengthens inference about cause-effect relationships in biofilm structural dynamics.

3.8 Linking Qualitative Observations to Quantitative Descriptors (Brief Overview)

While visual interpretation provides intuitive insights, many studies complement it with quantitative metrics extracted from CLSM Z-stacks using dedicated analysis software [5, 6].

Frequently used parameters include the following:

- *Bio volume*, defined as the total volume of biomass normalized to the imaged substratum area, reconstructed by counting above-threshold voxels representing cells or matrix.

- *Mean and maximum thickness*, determined by measuring the distance from the substratum to the biofilm surface across the field of view and summarizing these values statistically.
- *Surface coverage*, representing the proportion of the substratum area occupied by detectable biomass in a given XY plane.
- *Roughness coefficient*, a measure of spatial variation in thickness or height; higher roughness values indicate more heterogeneity and a more irregular surface profile (*see* **Note 10**).

Tools such as COMSTAT, COMSTAT2, and FIJI-based pipelines can automate or semi-automate these calculations, allowing robust comparisons between treatments, strains, and environmental conditions, and enabling correlations between architectural traits and outcomes such as antimicrobial susceptibility, mass transfer properties, and detachment behavior.

4 Notes

1. Use the lowest laser power and shortest exposure times that still yield acceptable image quality; rely on line or frame averaging to improve signal-to-noise instead of simply increasing laser intensity.
2. Limit the number of repeated scans of the same region, particularly in time-lapse experiments, to preserve cell viability and fluorescence [7–9].
3. Extend incubation times or gently agitate the sample during staining to encourage deeper penetration of the dyes.
4. Where compatible with the experimental aims, choose smaller fluorophore conjugates or adjust permeability conditions to improve entry without severely altering biofilm physiology.
5. Confirm correct coverslip thickness settings and proper objective immersion (no air bubbles).
6. Re-set the pinhole to the recommended value (~1 Airy unit) and verify that the Z-step size is appropriate for the objective and the sample.
7. Use unstained controls to identify intrinsic autofluorescence from the medium, plastic ware, or surface materials.
8. Optimize stain concentrations and washing steps to reduce nonspecific or excess fluorescence without stripping the biofilm.
9. Incorporate spacers or specialized imaging chambers to preserve the original thickness and prevent deformation of fragile or highly structured biofilms.
10. Handle coverslips gently and avoid placing heavy pressure when sealing the preparations.

Concluding Remarks

Confocal laser scanning microscopy has become a central technique for exploring biofilm architecture because it allows nondestructive, three-dimensional visualization of living communities in their hydrated state. When combined with thoughtful experimental design, appropriate fluorescent labeling strategies, carefully optimized acquisition parameters, and systematic observation frameworks, CLSM offers a powerful bridge between qualitative structural insight and quantitative analysis. Integrating these imaging-based descriptions with functional assays and modeling provides a deeper understanding of how spatial organization contributes to the robustness, adaptability, and pathogenic or beneficial roles of biofilms in diverse environments.

Competing Interest

The authors declare that they have no financial competing interests.

References

1. Arias-Méndez J, Gonzalez-Avalos J, Perera-Costa D, Vega-Hernandez W, Chávez-Santoscoy RA (2021) Visualisation and biovolume quantification in the Pseudomonas fluorescens biofilm using confocal laser scanning microscopy. Sci Rep 11(1):10405. https://doi.org/10.1038/s41598-021-90455-5
2. Mhade S, Rani SA, Kumar N, Karthik L (2023) Tools of the trade: image analysis programs for confocal micrographs of biofilms. ACS Omega 8(22):19180–19192. https://doi.org/10.1021/acsomega.2c07255
3. Alhede M, Qvortrup K, Liebrechts R, Østergaard C, Kragh KN, Nielsen PH et al (2012) Combination of microscopic techniques reveals a comprehensive biofilm matrix architecture in Pseudomonas aeruginosa infections. FEMS Immunology & Medical Microbiology 65(2):335–342. https://doi.org/10.1111/j.1574-695X.2012.00940.x
4. Merod RT, Mah TF, Chen DT (2007) Toward automated analysis of biofilm architecture: bias and variance. Appl Environ Microbiol 73(15): 4826–4834. https://doi.org/10.1128/AEM.00023-07
5. Reichhardt C, Parsek MR (2019) Confocal laser scanning Microscopy for analysis of Pseudomonas aeruginosa biofilm architecture and matrix localization. Front Microbiol 10:677. https://doi.org/10.3389/fmicb.2019.00677
6. Neu TR, Lawrence JR (2008) Biofilm analysis by confocal microscopy—basics and advanced approaches. Microsc Res Tech 76(2):162–177. https://doi.org/10.1002/jemt.70064
7. Relucenti M, Donfrancesco O, Familiari G, Cultura F, Rossi R, Taurino M, Di Giulio M (2021) Microscopy methods for biofilm imaging: focus on SEM. Int J Mol Sci 22(1):502. https://www.ncbi.nlm.nih.gov/pmc/articles/PMC7828176/
8. Confocal Laser Scanning Microscopy (n.d.) In J. Wang (Ed.), ScienceDirect topics. Elsevier. https://www.sciencedirect.com/topics/chemistry/confocal-laser-scanning-microscopy
9. Rani SA, Kono HM (2023) Visualization of biofilms using confocal laser scanning Microscopy. Bio-Protocol. https://bio-protocol.org/exchange/minidetail?id=4750315&type=30

Chapter 5

Detection and Structural Characterization of Microbial Biofilms Using Scanning Electron Microscopy

Priya P K, Vinushiya S, and Abirami K

Abstract

Biofilms represent structured microbial communities that attach to biotic or abiotic surfaces and become embedded within a self-produced extracellular matrix. Their formation is associated with increased resistance to antimicrobial agents and environmental stress, making accurate structural characterization essential for microbiological and biomedical research. This chapter focuses on the detection and morphological evaluation of biofilms using scanning electron microscopy (SEM). Biofilms were cultivated on selected substrate surfaces under controlled laboratory conditions, followed by systematic preparation involving fixation, graded dehydration, drying, and conductive coating to preserve structural integrity. SEM analysis enabled high-resolution visualization of surface-adhered microbial cells, microcolony formation, and extracellular polymeric matrix organization. Distinct developmental stages, including initial attachment, cluster formation, and mature multilayered architecture, were clearly observed.

The technique provided detailed insight into surface topography and biofilm–material interactions, demonstrating its effectiveness in assessing structural characteristics and treatment-induced alterations. Although SEM does not differentiate viable from non-viable cells and requires extensive sample preparation, it remains a powerful tool for examining biofilm morphology at high magnification. Overall, SEM-based biofilm detection offers reliable structural evidence of microbial colonization and serves as an important complementary method in biofilm research, particularly in studies involving material surfaces and antimicrobial evaluation.

Key words Biofilm architecture, Surface morphology, Microbial adhesion, Biofilm formation, Fixation and dehydration, Sputter coating, High-resolution imaging

1 Introduction

Microorganisms rarely exist as isolated, free-floating cells in natural or clinical environments. Instead, they predominantly grow as organized communities known as biofilms. These structured microbial assemblies adhere to living tissues, medical devices, industrial pipelines, and a wide variety of natural and artificial surfaces. Within a biofilm, cells are embedded in a self-produced extracellular polymeric matrix that provides mechanical stability

Abirami K (ed.), *Microbial Biofilm Research: Comprehensive Protocols and Best Practices*, Springer Protocols Handbooks, https://doi.org/10.1007/978-1-0716-5535-1_5,

and protection against environmental stress. This mode of growth significantly alters microbial behavior, including nutrient utilization, gene expression, and susceptibility to antimicrobial agents [1–3].

The formation of biofilms has important implications in medicine, industry, and environmental systems. In clinical settings, biofilms are associated with persistent infections, particularly those involving indwelling medical devices such as catheters, prosthetic implants, and contact lenses. In industrial contexts, biofilm accumulation can lead to biofouling, corrosion, and contamination of water systems. Because biofilm-associated microorganisms exhibit enhanced resistance to antibiotics and disinfectants, understanding their structural organization is essential for developing effective control strategies. Accurate detection and visualization of biofilms play a critical role in biofilm research. While conventional microbiological methods can quantify microbial growth, they do not provide detailed information about spatial organization, surface attachment, or extracellular matrix distribution. Microscopy-based techniques have therefore become indispensable tools for studying biofilm architecture. Among these techniques, scanning electron microscopy (SEM) has gained considerable importance due to its ability to provide high-resolution images of surface morphology.

SEM enables detailed examination of microbial attachment patterns, microcolony formation, and the complex three-dimensional arrangement of cells within the biofilm matrix. By scanning the specimen surface with a focused beam of electrons, SEM generates images that reveal topographical features at nanometer-scale resolution. This makes it particularly suitable for investigating biofilm–surface interactions and structural alterations following antimicrobial treatment. However, because biofilms are highly hydrated and structurally delicate, careful sample preparation is required prior to SEM analysis. Processes such as fixation, dehydration, drying, and conductive coating are essential to preserve morphological integrity and prevent imaging artifacts. When performed correctly, SEM provides valuable insights into biofilm development stages, structural heterogeneity, and treatment-induced changes [4].

This chapter focuses on the detection and structural characterization of biofilms using scanning electron microscopy. Emphasis is placed on systematic sample preparation, imaging parameters, and interpretation of morphological features. By employing SEM as a primary analytical tool, this work aims to contribute to a clearer understanding of biofilm architecture and its relevance in biomedical and material science research.

2 Materials

The materials used in this study were selected to support the cultivation, preservation, and ultrastructural examination of microbial biofilms using scanning electron microscopy (SEM). All reagents were of analytical grade, and procedures were conducted under aseptic conditions to ensure experimental accuracy and reproducibility.

2.1 Microbial Strain

A clinically or environmentally relevant bacterial strain was used for biofilm formation studies. The organism was maintained on appropriate agar media and subcultured in nutrient broth prior to experimentation to ensure viability and purity.

2.2 Growth Media

- Nutrient broth or other organism-specific culture medium.
- Nutrient agar plates for maintenance and subculturing.
- Phosphate-buffered saline (PBS), pH 7.2–7.4.

  The growth medium was prepared according to the manufacturer's instructions and sterilized by autoclaving at 121 °C for 15–20 min.

  (a) *Substrate Materials*

  - Sterile glass coverslips (18 × 18 mm or equivalent).
  - Stainless steel coupons or polymeric material (depending on experimental objective).
  - 24-well sterile tissue culture plates.

  Substrates were selected based on their relevance to the study (e.g., biomedical, industrial, or environmental applications).

  (b) *Fixatives and Chemicals*

  - Glutaraldehyde (2.5% in PBS)—primary fixative.
  - Osmium tetroxide (1%)—post-fixative (optional but recommended).
  - Ethanol solutions (30%, 50%, 70%, 80%, 90%, 95%, and 100%).
  - Hexamethyldisilazane (HMDS)—optional drying agent.
  - Distilled water.

  All chemical reagents were prepared freshly or stored according to manufacturer guidelines to maintain effectiveness.

(c) *Equipment*

- Incubator (temperature-controlled).
- Laminar airflow cabinet.
- Micropipettes and sterile tips.
- Centrifuge.
- Forceps (sterile).
- Critical point dryer (if available).
- Sputter coater (gold or gold–palladium target).
- Aluminum SEM stubs.
- Carbon adhesive tabs.
- Scanning electron microscope.

(d) *Personal Protective Equipment*

- Laboratory coat.
- Gloves.
- Protective eyewear.
- Fume hood access (for handling osmium tetroxide and HMDS).

3 Methods

A carefully controlled sample preparation procedure is essential for preserving the structural integrity of biofilms prior to SEM analysis (Table 1). Because biofilms are delicate, hydrated biological assemblies, improper handling can easily lead to distortion, shrinkage, or detachment from the substrate. The following stepwise protocol outlines each stage in detail, from substrate preparation to imaging [5–7].

3.1 Preparation of Substrates

1. The selection and preparation of substrates represent the first critical step in biofilm studies. Commonly used substrates include glass coverslips, stainless steel coupons, polymeric materials, or medical-grade implants, depending on the experimental objective.
2. Initially, substrates must be thoroughly cleaned to remove surface contaminants that may interfere with microbial adhesion. Cleaning may be performed using detergent washing followed by rinsing with distilled water. After cleaning, sterilization is carried out either by autoclaving at 121 °C for 15–20 min or by immersion in 70% ethanol followed by air drying under sterile conditions.

Table 1
Comparative overview of major microscopy techniques used in biofilm analysis, including scanning electron microscopy (SEM), confocal laser scanning microscopy (CLSM), and transmission electron microscopy (TEM)

Parameter	Scanning electron microscopy (SEM)	Confocal laser scanning microscopy (CLSM)	Transmission electron microscopy (TEM)
Basic principle	Uses a focused electron beam to scan the surface of a specimen and detect secondary electrons for high-resolution surface imaging	Uses laser excitation and optical sectioning to generate fluorescence-based three-dimensional images of biofilms	Transmits an electron beam through ultra-thin sections of a specimen to visualize internal ultrastructure
Type of imaging	Surface topography and external morphology	Three-dimensional architecture and spatial organization	Internal cellular and subcellular structure
Resolution	High resolution (nanometer-scale surface detail)	Moderate resolution (limited by light diffraction)	Extremely high resolution (ultrastructural detail at nanometer level)
Sample state	Dehydrated, fixed, and metal-coated samples	Hydrated samples; can image live or fixed biofilms	Fixed, dehydrated, ultra-thin sectioned samples
Preparation complexity	Extensive preparation including fixation, dehydration, drying, and sputter coating	Minimal to moderate preparation; staining with fluorescent dyes required	Highly complex preparation involving fixation, dehydration, embedding, sectioning, and staining
Visualization of EPS matrix	Visible as surface-associated matrix but limited chemical specificity	Can selectively stain EPS components using fluorescent probes	Limited; mainly structural visualization rather than matrix composition
Live/dead differentiation	Not possible	Possible using viability stains (e.g., live/dead staining)	Not possible
3D reconstruction	Limited (pseudo-3D appearance only)	True 3D reconstruction through optical sectioning	No 3D imaging; provides 2D internal sections
Biofilm thickness measurement	Indirect estimation from cross-sectional preparation	Accurate measurement through Z-stack imaging	Possible in cross-section but technically demanding
Observation of cell–surface interaction	Excellent for analyzing attachment to material surfaces	Moderate; depends on staining and optical clarity	Limited unless specifically sectioned at interface
Observation of intracellular details	Not visible	Not visible	Excellent visualization of organelles and intracellular structures

(continued)

Table 1 (continued)

Parameter	Scanning electron microscopy (SEM)	Confocal laser scanning microscopy (CLSM)	Transmission electron microscopy (TEM)
Artifacts and limitations	Dehydration may cause shrinkage and structural collapse	Photobleaching and limited penetration depth in thick biofilms	Sectioning artifacts and high technical demand
Time requirement	Moderate to high	Relatively faster once staining is completed	Very time-consuming
Cost and equipment requirement	Expensive instrumentation and maintenance	Expensive but widely available in research laboratories	Very high cost and specialized expertise required
Suitability for quantitative analysis	Semi-quantitative (surface coverage estimation)	Highly suitable for quantitative structural and viability analysis	Limited quantitative application in whole biofilm assessment
Typical application in biofilm research	Surface morphology, structural integrity, antimicrobial effect assessment	Biofilm architecture, viability studies, EPS distribution, thickness analysis	Ultrastructural analysis of cells within biofilm

This comparison highlights the strengths and limitations of each technique for structural and functional biofilm characterization

3. Sterile substrates are then aseptically transferred into the wells of a sterile 24-well tissue culture plate using sterile forceps. Care must be taken to avoid scratching the surface, as physical damage may alter biofilm attachment patterns. Once placed in the wells, substrates are completely submerged in sterile growth medium to ensure uniform exposure during inoculation.
4. Proper substrate preparation ensures reproducibility and prevents experimental variability arising from surface contamination or inconsistent sterilization (Fig. 1).

3.2 Biofilm Formation

1. Biofilm development begins with the preparation of a standardized bacterial inoculum. A fresh overnight culture grown in an appropriate nutrient broth is typically used. The culture turbidity is adjusted to a defined optical density (commonly OD600 $\approx$ 0.1), ensuring uniform initial cell concentration across experimental replicates.
2. A measured volume of the standardized inoculum is added to each well containing the prepared substrates. The plates are then incubated under conditions optimal for the test organism, such as 35–37 °C for many human pathogens or other temperature ranges depending on the species. Atmospheric conditions (aerobic, microaerophilic, or anaerobic) must be controlled as required.

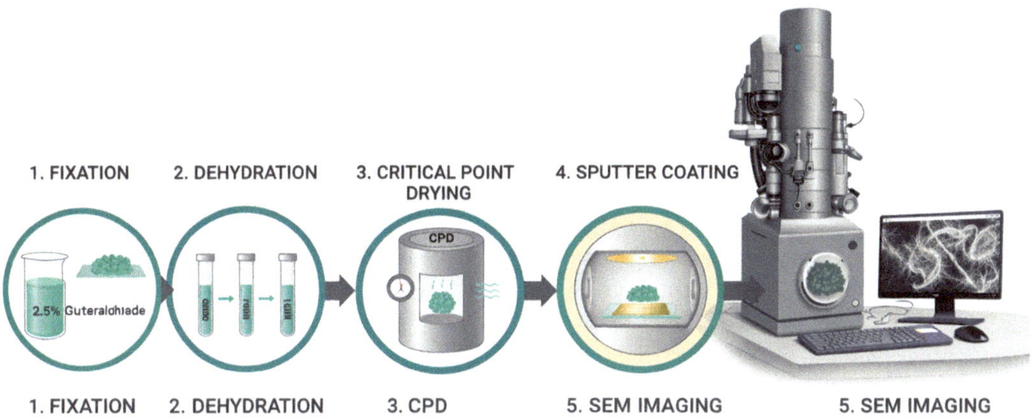

Fig. 1 Systematic workflow involved in SEM-based biofilm analysis, beginning with substrate preparation and sterilization, followed by controlled biofilm cultivation under defined incubation conditions

3. Biofilm development is allowed to proceed for predetermined time intervals (e.g., 24, 48, or 72 h). These time points allow evaluation of early adhesion, intermediate microcolony formation, and mature biofilm architecture. Unless shear stress is being specifically studied, incubation is performed under static conditions to promote surface attachment (*see* **Note 1**).
4. It is important to avoid unnecessary disturbance during incubation, as agitation may dislodge loosely attached cells and influence biofilm maturation patterns.

3.3 Washing Step

1. Following incubation, the removal of non-adherent planktonic cells is necessary to ensure that only surface-associated biofilm structures are examined.
2. The culture medium is gently aspirated using a micropipette without touching the substrate surface. This step requires precision, as direct contact may disrupt fragile biofilm layers. The substrates are then carefully rinsed two to three times with sterile phosphate-buffered saline (PBS). Each wash should be performed slowly along the wall of the well rather than directly over the biofilm.
3. The purpose of washing is to eliminate loosely attached cells while maintaining the integrity of firmly adhered microbial communities. Excessive force during washing may result in artificial reduction of biofilm density, leading to inaccurate interpretation of results.

3.4 Primary Fixation

4. Fixation stabilizes biofilm architecture by preserving cellular morphology and extracellular matrix components.
5. Substrates bearing biofilms are immersed in 2.5% glutaraldehyde prepared in PBS. Glutaraldehyde acts as a cross-linking

agent that binds to amino groups in proteins, thereby stabilizing cellular structures and preventing autolysis. Samples are incubated in the fixative solution for 2–4 h at 4 °C. Low-temperature fixation reduces enzymatic degradation and enhances structural preservation.

6. After fixation, samples are gently rinsed two to three times with PBS to remove residual fixative. Proper fixation is crucial, as inadequate fixation may result in collapsed or distorted cell morphology during subsequent dehydration steps.

3.5 Post-Fixation (Optional but Recommended)

1. Post-fixation with 1% osmium tetroxide further enhances structural stability, particularly of lipid-containing membranes.
2. Osmium tetroxide reacts with unsaturated lipids and provides improved membrane contrast during SEM imaging. Samples are typically exposed to osmium tetroxide for 1–2 h at room temperature. Because osmium tetroxide is volatile and toxic, this step must be performed inside a chemical fume hood with appropriate protective equipment.
3. Following treatment, specimens are rinsed thoroughly with distilled water to remove any remaining osmium residues. Although optional, this step significantly improves image quality and structural preservation (*see* **Note 2**).

3.6 Dehydration Series

1. Since SEM requires a vacuum environment, all water must be removed from biological samples. Gradual dehydration minimizes structural damage caused by rapid water loss.
2. Samples are sequentially immersed in increasing concentrations of ethanol:
   - 30% ethanol for 10 min.
   - 50% ethanol for 10 min.
   - 70% ethanol for 10 min.
   - 80% ethanol for 10 min.
   - 90% ethanol for 10 min.
   - 95% ethanol for 10 min.
   - 100% ethanol for 15 min (repeated twice).

This graded series allows controlled replacement of water with ethanol, reducing osmotic stress and shrinkage. Complete dehydration is essential, as residual moisture may lead to biofilm collapse under vacuum conditions.

3.7 Drying

Drying converts the dehydrated sample into a stable form suitable for SEM examination.

(a) Critical Point Drying (Preferred Method).

- Critical point drying (CPD) is widely regarded as the most effective technique for preserving biofilm architecture. In this method, ethanol within the sample is replaced with liquid carbon dioxide.
- The specimen is then brought to the critical point of CO_2, where liquid and gas phases are indistinguishable.
- This prevents surface tension forces that can distort delicate biofilm structures during drying.

(b) Hexamethyldisilane (HMDS) Drying (Alternative Method).

- As a cost-effective alternative, the final ethanol is replaced with hexamethyldisilane (HMDS). The sample is then allowed to air dry in a fume hood.
- Although simpler than CPD, HMDS drying may produce minor shrinkage artifacts but generally preserves morphology adequately for routine analysis.
- Critical point drying remains preferable for high-resolution structural studies (*see* **Note 3**).

3.8 Mounting and Sputter Coating

1. Dried samples are mounted onto aluminum SEM stubs using conductive carbon adhesive tabs. Proper mounting ensures sample stability during imaging.
2. Because biological samples are non-conductive, they must be coated with a thin conductive metal layer to prevent electron charging. Sputter coating with gold or gold–palladium is commonly performed for 30–60 s, producing a uniform conductive film (*see* **Note 4**).
3. Adequate coating improves signal detection and image sharpness while minimizing charging artifacts [8–10].

3.9 SEM Imaging

1. Prepared samples are placed into the SEM chamber, and vacuum conditions are established. Imaging parameters are adjusted based on specimen characteristics.
2. An accelerating voltage between 5 and 20 kV is typically used. Lower voltages are preferred for delicate biological specimens to reduce beam damage and improve surface detail. Images are captured at various magnifications, often ranging from 1000× to 20,000× or higher, depending on the required structural resolution.
3. Multiple fields from different regions of the substrate should be recorded to ensure representative analysis. Consistency in imaging parameters across experimental groups is essential for comparative studies (*see* **Note 5**).

3.10 Interpretation of SEM Results

SEM analysis provides detailed visualization of biofilm architecture at different developmental stages.

1. Early-stage biofilms typically show individual cells attached sporadically to the substrate surface. As biofilm development progresses, cells form microcolonies characterized by clustered bacterial aggregates. Mature biofilms appear as dense, multilayered structures embedded in a complex extracellular polymeric matrix, often exhibiting heterogeneous surface topography.
2. The extracellular polymeric substance (EPS) may appear as a fibrous or web-like network surrounding bacterial cells. In antimicrobial treatment studies, SEM can reveal disrupted architecture, reduced cell density, surface pitting, or damaged cell morphology.
3. Quantitative comparisons may include estimation of surface coverage, morphological changes, and biofilm thickness patterns.

3.11 Advantages of SEM in Biofilm Research

SEM offers several significant advantages:

- High-resolution visualization of surface morphology.
- Three-dimensional appearance of biofilm topography.
- Clear observation of microbial attachment and clustering.
- Effective analysis of material–biofilm interactions.
- Comparative structural evaluation before and after treatment.

These advantages make SEM a valuable tool in microbiology, material science, and biomedical research.

3.12 Limitations

Despite its usefulness, SEM has certain limitations:

- Extensive and time-consuming sample preparation.
- Potential for dehydration-induced shrinkage artifacts.
- Inability to distinguish viable from non-viable cells.
- Visualization limited primarily to surface features.

Therefore, SEM is often complemented with other techniques such as fluorescence microscopy for viability assessment.

3.13 Precautions

To ensure reliable results, the following precautions should be observed:

- Osmium tetroxide must be handled in a certified fume hood due to its toxicity.
- Washing steps should be gentle to prevent mechanical detachment of biofilm layers.
- Complete dehydration is essential before drying.

- Electron beam exposure should be minimized to prevent structural damage.
- Consistency in fixation and imaging parameters should be maintained across experiments.

4 Notes

1. Poor or Inconsistent Biofilm Formation

Problem

Minimal cell attachment or uneven biofilm distribution on substrates.

Possible Causes

- Low initial inoculum concentration
- Inappropriate incubation conditions (temperature, time, atmosphere)
- Incompatible substrate surface properties
- Excessive disturbance during incubation

Solutions

- Standardize inoculum density (e.g., consistent optical density before inoculation).
- Optimize incubation time to capture early, intermediate, and mature biofilm stages.
- Pre-condition substrates if necessary (e.g., surface treatment to enhance adhesion).
- Maintain static incubation unless shear conditions are being studied.

Consistency in experimental design is essential for reproducible biofilm formation.

2. Biofilm Detachment During Washing

Problem

Significant loss of attached biofilm during PBS washing steps.

Possible Causes

- Vigorous pipetting or direct fluid flow onto biofilm surface
- Insufficient maturation time
- Weak surface adhesion

Solutions

- Perform gentle washing along the wall of the well rather than directly over the sample.
- Use minimal washing cycles (2–3 gentle rinses).
- Allow sufficient incubation time for stable attachment.
- Handle substrates carefully using fine sterile forceps.

Excessive mechanical force is one of the most common causes of structural disruption.

3. Structural Collapse or Shrinkage After Dehydration

Problem
Biofilm appears flattened, cracked, or distorted in SEM images.

Possible Causes

- Rapid dehydration
- Incomplete fixation
- Improper drying method

Solutions

- Follow a gradual ethanol series without skipping concentrations.
- Ensure adequate fixation time with fresh fixative.
- Prefer critical point drying over air drying when possible.
- Avoid prolonged exposure to high temperatures during drying.

Gradual dehydration and controlled drying are critical for maintaining biofilm architecture.

4. Charging Artifacts During Imaging

Problem

Bright spots, image drift, or distorted areas appearing during SEM imaging.

Possible Causes

- Inadequate conductive coating
- Insufficient sputter coating thickness
- Poor grounding of the sample

Solutions

- Ensure uniform gold or gold–palladium coating.
- Increase sputter coating duration slightly if charging persists.
- Confirm proper contact between sample and stub using conductive adhesive.

Non-conductive biological samples are highly prone to charging; proper coating is essential.

5. Low Image Contrast or Blurred Images

Problem
Poor resolution or lack of surface detail.
Possible Causes

- Improper accelerating voltage
- Contaminated sample surface
- Beam damage due to prolonged exposure

Solutions

- Optimize accelerating voltage (often 5–10 kV for biological samples).
- Ensure complete dehydration before imaging.
- Minimize beam exposure time.
- Clean SEM chamber if contamination is suspected.

Fine adjustment of instrument parameters significantly improves imaging clarity.

Competing Interest

The authors state that they have no financial interests that could be perceived as influencing the content of this chapter.

References

1. Chakraborty A, Dam B (2025) A rapid and efficient method for visualisation of microbial biofilms on natural, and industrially- and medically-relevant surfaces, using field emission-scanning electron microscopy. J Microbiol Methods 239:107315. https://doi.org/10.1016/j.mimet.2025.107315
2. Wells M, Mikesh M, Gordon V (2024) Structure-preserving fixation allows scanning electron microscopy to reveal biofilm microstructure and interactions with immune cells. J Microsc 293(1):59–68. https://doi.org/10.1111/jmi.13252
3. Liu H, Wang X, Wang Z, Shen Y (2025) Evaluation of bacterial biofilm, smear layer, and debris removal efficacy ...: a CLSM and SEM study. BMC Oral Health 25:95. https://doi.org/10.1186/s12903-025-05463-y
4. Rimal B et al (2024) Scanning electron microscopy and energy-dispersive X-ray spectroscopy of Staphylococcus aureus biofilms. ACS

Omega 9(36):37610–37620. https://doi.org/10.1021/acsomega.4c01168

5. Zmerli M et al (2025) Comparison of tabletop SEM, conventional SEM, and TEM relevant to diagnostic potential. Front Cell Infect Microbiol. https://doi.org/10.3389/fcimb.2025.1697696
6. Kozlov KS, Boiko DA, Detusheva EV, Detushev KV, Pentsak EO, Vereshchagin AN, Ananikov VP (2023) Digital biology approach for macroscale studies of biofilm growth and biocide effects with electron microscopy. Digital Discov 2:1522–1539. https://doi.org/10.1039/D3DD00048F
7. Ratnayake K, Joyce DC, Webb RI (2012) A convenient sample preparation protocol.... Sci Hortic 140:12–18. https://doi.org/10.1016/j.scienta.2012.03.012
8. Bossù M, Selan L, Artini M, Relucenti M, Familiari G, Papa R, Vrenna G, Spigaglia P, Barbanti F, Salucci A, Di Giorgio G, Rau JV, Polimeni A (2020) Characterization of *Scardovia wiggsiae* biofilm by original scanning electron microscopy protocol. Microorganisms 8(6):807. https://doi.org/10.3390/microorganisms8060807
9. Relucenti M, Familiari G, Donfrancesco O, Taurino M, Li X, Chen R, Artini M, Papa R, Selan L. Microscopy methods for biofilm imaging: focus on SEM and VP-SEM pros and cons. Biology (Basel) 2021;10(1):51. https://doi.org/10.3390/biology10010051
10. Chakraborty A, Dam B (2025) A rapid and efficient method for visualisation of microbial biofilms on natural, and industrially- and medically-relevant surfaces, using field emission- scanning electron microscopy. J Microbiol Methods 239:107315. https://doi.org/10.1016/j.mimet.2025.107315

Chapter 6

Detection and Ultrastructural Characterization of Biofilms by Transmission Electron Microscopy

Abirami J and Abirami K

Abstract

Bacteria predominantly exist in the form of biofilms, which play a critical role across numerous industrial, medical, and agricultural sectors. However, the analysis of biofilms remains challenging due to their complex composition, structural heterogeneity, and dynamic nature. The spectroscopy technique is a versatile analytical technique capable of examining a wide range of samples, including solids, liquids, and gases, and is widely applied in disciplines such as chemistry, physics, materials science, and biology. However, transmission electron microscopy (TEM) employs a focused beam of high-energy electrons to investigate ultrathin specimens, typically less than 200 nm in thickness, enabling visualization at the nanoscale. When combined with advanced techniques such as precession-based three-dimensional electron diffraction tomography, TEM has demonstrated significant potential in crystal structure determination. A key advantage of this approach is its ability to analyze extremely small single crystals, ranging from 25 to 500 nm, while requiring only minimal amounts of sample material.

Key words Electron diffraction tomography, Crystal structure identification, Nanoscale imaging, Microbial characterization, Analytical techniques

1 Introduction

Biofilms can develop on a wide range of surfaces, including biotic surfaces such as host tissues, teeth, and mucosal membranes, as well as abiotic surfaces like medical implants, catheters, industrial pipelines, and ship hulls. They are ubiquitous in natural, industrial, and clinical environments. Structurally, biofilms are highly heterogeneous and contain complex architectures with interconnected water channels that facilitate the transport of nutrients, signaling molecules, and metabolic waste products. Microorganisms within biofilms exhibit phenotypic characteristics that are markedly different from their free-living (planktonic) counterparts. Biofilm-associated cells demonstrate increased tolerance to antimicrobial agents, host immune defense, and environmental stresses. This enhanced resistance arises from multiple factors, including the

Abirami K (ed.), *Microbial Biofilm Research: Comprehensive Protocols and Best Practices*, Springer Protocols Handbooks, https://doi.org/10.1007/978-1-0716-5535-1_6,

restricted penetration of antimicrobial compounds through the EPS matrix, the presence of e-DNA and extracellular enzymes such as β-lactamases and toxins, altered metabolic activity, limited availability of oxygen and nutrients, and the existence of persister cell populations [1].

The initiation, development, and maturation of biofilms are regulated by microbial surface proteins and polysaccharide intercellular adhesins (PIA), which promote cell-to-cell and cell-to-surface interactions. As a result, infections involving biofilm-embedded microorganisms are significantly more difficult to eradicate than those caused by planktonic cells. Biofilms are implicated in a wide range of chronic and device-associated infections, including endocarditis, urinary tract infections, septic arthritis, chronic rhinosinusitis, ocular infections, wound infections, and infections related to indwelling medical devices. Biofilm formation on such devices often leads to persistent, recurrent infections and may ultimately result in device malfunction or failure [2].

2 Materials

2.1 Sample Collection and Initial Handling

- Sterile collection tools (swabs, scrapers, scalpels) and sterile containers.
- Primary sample buffer (e.g., physiological buffer appropriate for your organism).
- Cold storage (ice, refrigerated transport) and labelling supplies.

2.2 Fixation and Buffering (to Preserve Ultrastructure)

- Primary chemical fixative (commonly glutaraldehyde or paraformaldehyde—preserves proteins).
- Secondary post fixative (commonly osmium tetroxide—stabilizes membranes; very *toxic*; institutional controls required).
- Buffer system (e.g., sodium cacodylate buffer or phosphate buffer)—used to maintain pH during fixation.
- Quenching/washing solutions (same buffer).

Safety note: osmium tetroxide and some buffers are highly hazardous. Follow institutional safety procedures and waste disposal rules.

2.3 Dehydration and Infiltration (for Embedded Thin-Section TEM)

- Dehydrating solvents: graded ethanol or acetone series (progressive steps).
- Embedding resins: epoxy resins (e.g., Epon, Araldite) or other TEM-grade resins.
- Resin curing oven (for polymerization).

2.4 Sectioning and Grid Preparation

- Ultramicrotome capable of producing ultrathin sections (~50–90 nm).
- Diamond knife (ultramicrotomy) and boat.
- TEM grids (copper, nickel, or gold; 200–400 mesh commonly used).
- Support films: Formvar and/or carbon-coated grids (or pre-coated grids).
- Grid storage boxes and fine forceps/tweezers.

2.5 Staining (Contrast Enhancement)

- Heavy-metal stains for sections: uranyl acetate (widely used) and lead citrate (commonly used).
- For whole-mount/negative staining of cells/extracellular polymeric substances (EPS): phosphotungstic acid (PTA) or uranyl acetate alternatives.
- Distilled/deionized water and filtered staining receptacles.

  Safety/regulatory note: uranyl compounds contain uranium and are regulated/toxic; institutional approval and radiation safety procedures may apply. Lead compounds are toxic.

2.6 Instrumentation and TEM Accessories

- Transmission electron microscope (suitable accelerating voltage for biological specimens).
- TEM specimen holder(s) compatible with grids.
- Glow discharger or plasma cleaner (to make grids hydrophilic when needed).
- Carbon evaporator/sputter coater (optional—for additional carbon coating of grids).
- Grid inspection microscope (light microscope for handling).

3 Methods

3.1 Sample Collection

1. Carefully collect biofilm samples from the surface (e.g., glass, medical device, or tissue) using sterile forceps or a scalpel [3, 4].
2. Minimize mechanical disruption to preserve the native biofilm structure.
3. Immediately immerse samples in primary fixative (Fig. 1).

3.2 Primary Fixation

1. Immediately after collection, biofilm samples are subjected to primary fixation to preserve cellular morphology and stabilize the extracellular polymeric substance (EPS) matrix. Samples are immersed in 2.5% glutaraldehyde prepared in 0.1 M phosphate buffer adjusted to a physiological pH of 7.2–7.4 Glutaraldehyde acts by cross-linking proteins, thereby maintaining the structural integrity of microbial cells and the surrounding biofilm matrix (*see* **Note 1**).

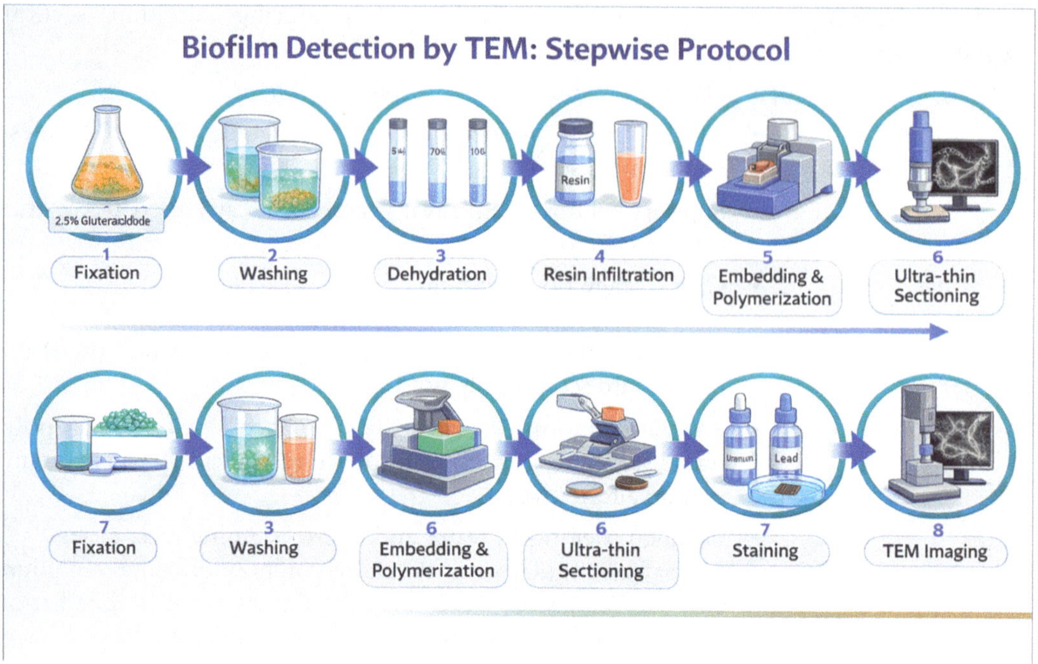

Fig. 1 Stepwise protocol for biofilm detection by transmission electron microscopy (TEM). The schematic illustrates the sequential workflow for ultrastructural analysis of biofilms using TEM. (1) Primary fixation: Biofilm samples are fixed in 2.5% glutaraldehyde to preserve cellular and extracellular matrix architecture. (2) Washing: Samples are rinsed in buffer to remove excess fixative. (3) Dehydration: Gradual dehydration is performed using increasing concentrations of ethanol or acetone. (4) Resin infiltration: Dehydrated samples are infiltrated with epoxy resin to provide structural support. (5) Embedding and polymerization: Samples are embedded in resin blocks and polymerized to harden. (6) Ultra-thin sectioning: Sections (~60–90 nm) are cut using an ultramicrotome and mounted on copper grids. (7) Contrast staining: Sections are stained with heavy metals (e.g., uranyl acetate and lead citrate) to enhance contrast. (8) TEM imaging: Prepared grids are examined under a transmission electron microscope to visualize biofilm ultrastructure, including bacterial cells and extracellular polymeric substances (EPS)

2. Fixation is carried out at 4 °C to reduce enzymatic activity and prevent autolytic degradation. Samples are incubated for 2–4 h, while thicker or more mature biofilms may require overnight fixation to ensure thorough penetration of the fixative. Gentle handling during this step is essential to avoid disruption of the biofilm architecture.
3. Following fixation, samples are rinsed three times for 10 min each with fresh phosphate buffer to remove excess glutaraldehyde. This washing step is critical to minimize background staining and prevent interference with subsequent post-fixation and dehydration procedures [5, 6].

Optional: Add 0.05–0.1% ruthenium red to the fixative to enhance visualization of EPS.

3.3 Post-Fixation

1. Following primary fixation, the samples are subjected to post-fixation to enhance membrane preservation and improve electron contrast. Biofilm specimens are immersed in 1% osmium tetroxide prepared in phosphate or cacodylate buffer and incubated for 1–2 h at 4 °C This step stabilizes lipid components and provides increased contrast to cellular membranes and extracellular structures, facilitating detailed ultrastructural analysis under TEM.
2. After post-fixation, the samples are gently rinsed two to three times with the same buffer to remove excess osmium tetroxide, followed by a final wash with distilled water to eliminate residual buffer salts that may interfere with subsequent dehydration and embedding steps.
3. Safety Consideration: Osmium tetroxide is extremely volatile, corrosive, and toxic. All handling must be carried out in a properly functioning chemical fume hood, with appropriate personal protective equipment, including gloves, lab coat, and eye protection. Waste containing osmium tetroxide should be disposed of according to institutional safety guidelines.

3.4 Dehydration

Following fixation and post-fixation, samples must be carefully dehydrated to remove all water prior to resin infiltration. Dehydration is performed using a graded series of ethanol or acetone to minimize structural distortion and preserve biofilm integrity (*see* **Note 2**).

1. Transfer the fixed biofilm samples sequentially through increasing concentrations of ethanol (or acetone) as follows: 30%, 50%, 70%, 90%, and 100% (v/v).
2. Incubate the samples for 10–15 min at each concentration, ensuring complete immersion at all times.
3. Gently agitate the samples during each step to facilitate uniform solvent exchange while avoiding mechanical disruption of the biofilm matrix.
4. Replace each dehydration solution with fresh solvent at the next higher concentration, taking care to minimize carryover from the previous step.
5. After the initial 100% ethanol step, perform two additional changes in fresh 100% ethanol, each for 10–15 min, to ensure complete removal of residual water.
6. Maintain samples at room temperature during dehydration unless otherwise required by the sample type.

3.5 Infiltration and Embedding

- Infiltrate samples with a mixture of epoxy resin (Epon/Spurr's) and ethanol:
  - 1:1 ethanol: resin for 1 h.
  - 1:3 ethanol: resin for 2–3 h.

- Transfer samples to pure resin and incubate overnight.
- Embed samples in molds and polymerize at 60 °C for 24–48 h.

3.6 Ultrathin Sectioning

- Trim resin blocks and cut ultrathin sections (60–90 nm) using an ultramicrotome with a diamond knife.
- Collect sections onto copper or nickel TEM grids.

3.7 Contrast Staining

- Stain grids with 2% uranyl acetate for 10–15 min.
- Rinse with distilled water and blot gently.
- Counterstain with lead citrate for 5–10 min.
- Wash with CO_2-free distilled water and air-dry.

3.8 TEM Imaging

1. Examine sections using a TEM operated at 80–120 kV.
2. Capture images showing:
   - Biofilm architecture.
   - Bacterial morphology.
   - EPS matrix.
   - Cell-to-cell interactions and surface attachment.

3.9 Key Features to Detect

1. Dense microbial cell clusters.
2. The presence of matrix-forming extracellular polymeric compounds (EPS).
3. Microcolonies incorporated into EPS.
4. The interactions between cells and the cell surface that define the architecture of biofilms.
5. Comparative Observation—Biofilm mode (organized matrix-embedded communities) from planktonic (free-floating) cells observation.
6. Analysis—Analysis of Capture electron micrographs for structural analysis and quantification of biofilm thickness, density, and morphology.

4 Notes

1. Gradual dehydration is critical for preserving the ultrastructure of the extracellular polymeric substance (EPS).
2. Rapid dehydration or prolonged exposure to intermediate ethanol concentrations may result in biofilm shrinkage or collapse.

Competing Interest

The authors confirm that there are no financial competing interests associated with this chapter.

References

1. Almatroudi A (2025) Biofilm resilience: molecular mechanisms driving antibiotic resistance in clinical contexts. Biology 14(2):165. https://doi.org/10.3390/biology14020165
2. Ugwu CN, Ezeibe EN, Emencheta SC, Nwagwu CS, Ogbonna KO, Ejiofor CV, Onugwu AL, Berebon DP, Attama AA (2025) Biofilms: structure, resistance mechanism, emerging control strategies, and applications. RSC Pharmaceutics 2:1376–1407. https://doi.org/10.1039/D5PM00094G
3. Tizro P, Choi C, Khanlou N (2019) Sample preparation for transmission electron microscopy. Methods Mol Biol 1897:417–424. https://doi.org/10.1007/978-1-4939-8935-5_33
4. Zhang YR, Qiao HL, Ren LL, Wang R, Lu PF (2020) Sample preparation method of scanning and transmission electron microscope for the appendages of Woodboring beetle. J Vis Exp (156). https://doi.org/10.3791/59251
5. McCutcheon J, Southam G (2018) Advanced biofilm staining techniques for TEM and SEM in geomicrobiology: implications for visualizing EPS architecture, mineral nucleation, and microfossil generation. Chem Geol 498:115–127.,ISSN 0009-2541. https://doi.org/10.1016/j.chemgeo.2018.09.016
6. Dumančić E, Vojta L, Fulgosi H (2023) Beginners guide to sample preparation techniques for transmission electron microscopy. Period Biol 125(1-2):123–131. ISSN 0031-5362. https://doi.org/10.18054/pb.v125i1-2.25293

Chapter 7

Crystal Violet–Based Colorimetric Assay for Biofilm Evaluation

Prabha Nagaram

Abstract

Surface-associated microbial colonies known as biofilms can be found in natural, medical, and industrial environments. Typically, they are encased in an extracellular matrix that offers structural stability and protection. Inside these biofilms, microbes organize into macrocolonies and create fluid-filled channels, making them highly resistant to antibiotics and host immune defenses. To investigate early stages of biofilm development, the microtiter plate assay is commonly employed, particularly for organisms such as *Pseudomonas aeruginosa*. In this assay, the resulting biofilms display characteristics similar to mature biofilms, such as antibiotic tolerance. The method enables biofilm growth on the walls or base of microtiter plates and is well suited for high-throughput testing of multiple strains under different conditions. Quantification is most often carried out using crystal violet (CV) staining, although alternative dyes can also be applied. Because of its versatility, affordability, and ease of use, the microtiter plate assay has emerged as a key method for researching biofilms in a variety of bacteria.

Key words Biofilm, Crystal violet, Bacteria, Cell proliferation, Microtiter plates

1 Introduction

Biofilm development serves as a key survival mechanism for many bacteria and fungi, where cells are enclosed within a protective extracellular matrix made of organic polymers [1]. In fact, biofilm growth is likely the dominant lifestyle for microbes in most natural settings. Within a biofilm, microorganisms typically form a distinct structure, consisting of macrocolonies interspersed with fluid-filled channels. The microtiter plate assay is a valuable method for examining the initial steps of biofilm development and has been used mainly to study bacterial biofilms. In this assay, biofilms can develop along the walls and/or at the base of the wells. It is especially useful for assessing biofilm development by several strains under various growing circumstances because of its high-throughput capability. Numerous bacteria, including *Vibrio cholerae*, *Pseudomonas*, *Staphylococcus*, *Escherichia coli*, *Mycobacterium*, *Enterococcus*, and several

Abirami K (ed.), *Microbial Biofilm Research: Comprehensive Protocols and Best Practices*, Springer Protocols Handbooks, https://doi.org/10.1007/978-1-0716-5535-1_7,

fungi, have had their early biofilm growth studied using modified versions of this assay [2]. Crystal violet (CV) is commonly used to quantify biofilm biomass. Christensen and associates first presented the crystal violet microtiter plate assay in the 1980s [3]. Its simplicity, affordability, and adaptability have established the microtiter plate assay as a key method in biofilm research.

In microbiology and cell biology, crystal violet staining is a frequently used technique for identifying and quantifying cells, biofilms, and other structural elements. Utilizing crystal violet dye's affinity for biological components like DNA, cell walls, and membranes, this assay operates. The crystal violet assay is an essential technique in many research scenarios, offering crucial information regarding cellular condition and viability by generating a distinct and measurable signal of cell density or biofilm formation [4]. In mammalian cell cultures, crystal violet staining is widely used to evaluate cell viability and growth indirectly. Staining intensity declines when cells lose viability because they separate from the culture surface, indicating decreased cell survival. This technique provides a consistent way to apply crystal violet in various experimental settings, assisting in the production of precise and repeatable results [5].

2 Materials

2.1 Reagents

The materials required for crystal violet staining include crystal violet dye, methanol, ethanol, and phosphate-buffered saline (PBS).

2.2 Crystal Violet

This dye is responsible for staining the biofilm.

Crystal violet, also referred to as basic violet or gentian violet, is a positively charged dye commonly used in different staining techniques. It functions by attaching to negatively charged cellular components, including nucleic acids and proteins. Since the dye is supplied as a powder, it needs to be dissolved in an appropriate solution before application. A typical working concentration is 0.1% in PBS, although this may be adjusted depending on the specific experimental purpose.

Furthermore, reagents like methanol, ethanol, PBS, sodium citrate, and acetic acid are needed for the staining process (*see* **Notes 1** and **2**).

2.3 Solvent (Such as Ethanol or Methanol)

Serves to dissolve the crystal violet powder and formulate the staining solution.

2.4 Buffer (Such as PBS, Water, or Ammonium Oxalate)

Employed to dilute the crystal violet solution and remove excess or unbound dye.

2.4.1 Elution Buffer (for Example, 30% Acetic Acid in Water)

Utilized to dissolve the crystal violet bound to the biofilm, enabling its quantification.

2.5 Washing Solution (Such as PBS or Water)

Applied to eliminate planktonic cells and excess dye prior to the staining step.

Sterile Water: Utilized in solution preparation and for washing procedures.

2.6 Equipment

A microplate reader is highly useful for high-throughput screening in multi-well plates, allowing accurate analysis of numerous samples at once. In contrast, a spectrophotometer is a simple optical instrument designed to measure light absorbance at different wavelengths. Alongside these, basic cell culture tools—including culture plates, target cells, media, supplements, and antibiotics are essential—for maintaining cell growth under controlled conditions and for studying their behavior and responses.

2.7 Preparation of Crystal Violet Solution

Preparing the crystal violet solution is a crucial step in staining methods, as the dye needs to be properly dissolved to ensure effective binding to cells or biofilms.

2.7.1 Crystal Violet Dye Solution

Crystal violet staining solution is typically prepared by dissolving the dye in PBS or distilled water, with proper dissolution being essential to ensure accurate cell viability studies, as undissolved particles may interfere with uniform staining. While the concentration can be adjusted based on experimental requirements, a 0.1% solution is most commonly used. To prepare this, 1 g of crystal violet powder is dissolved in 100 ml of PBS and mixed thoroughly until fully dissolved. Since crystal violet is sensitive to light, the solution should be stored in a dark container.

To prepare a 0.1% crystal violet solution in PBS, 0.1 grams of crystal violet powder is carefully weighed and added to 100 ml of PBS. The solution must then be stirred thoroughly until the dye is completely dissolved, ensuring a uniform and reliable working solution. Because crystal violet is highly sensitive to light, prolonged exposure can lead to degradation or reduced staining efficiency; therefore, the prepared solution should always be stored in a dark or opaque container to maintain its stability.

This 0.1% crystal violet solution has broad applications across microbiology and cell biology. It is extensively used for staining cells in culture, allowing visualization of cellular morphology and viability. In biofilm research, it plays a critical role in biofilm assays, where the dye binds to the extracellular matrix and cells, enabling quantification of biofilm biomass.

Additionally, crystal violet serves as the principal dye in Gram staining, a fundamental technique in microbiology for differentiating between Gram-positive and Gram-negative bacteria. These wide-ranging uses highlight the importance of careful preparation and handling of crystal violet solutions to ensure consistency, accuracy, and reliability in experimental outcomes.

3 Methods

3.1 Growing a Biofilm

In Lauria Bertani (LB) medium, cultivate wild-type strain or its mutant strain (such as *Pseudomonas aeruginosa*) for a day.

For biofilm assays, the overnight culture needs to be diluted 1: 100 in fresh media. Pour 100 μL of the diluted culture into each well of a 96-well plate.

For quantitative investigation, four to eight replicate wells per treatment are frequently used.The microtiter plate should be incubated for 4–24 h at 37 °C (Fig. 1).

3.2 Biofilm Staining

- To get rid of the cells after incubation, turn the plate over and drain the liquid.
- Gently dip the plate in a small container of water (you might use the bottom of a P1000 pipette tip box, for example). Repeat the procedure after shaking out the water. By removing detached

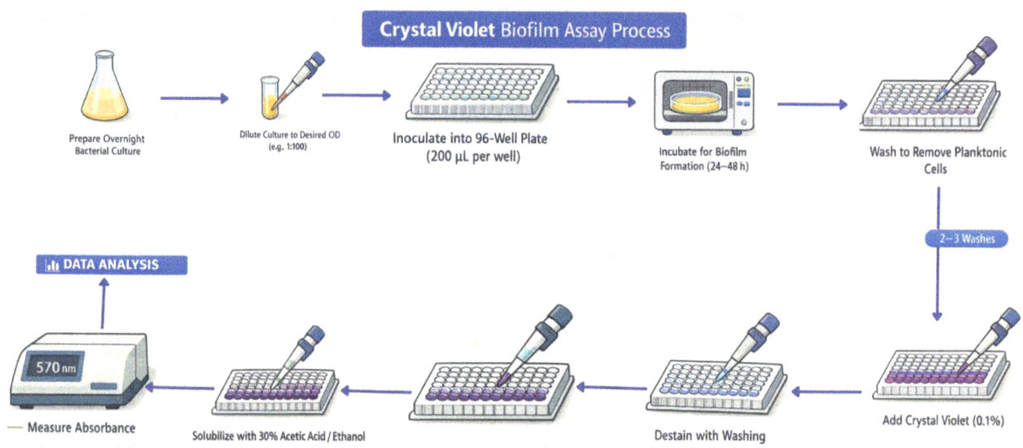

Fig. 1 Illustrates the step-by-step protocol for the biofilm assay. Overnight bacterial cultures are adjusted to the required optical density and inoculated into 96-well microtiter plates. After static incubation for 24–48 h to allow biofilm formation, non-adherent cells were removed and wells are gently washed with PBS. Adherent biofilms are fixed (if applicable), stained with 0.1% crystal violet, and excess stain is rinsed off. The bound dye is solubilized with ethanol or acetic acid, and biofilm biomass was quantified by measuring absorbance at 570–600 nm

cells and any remaining medium that can obstruct staining, this process lessens background staining.

- Leave the microtiter plate 10–15 min to incubate at room temperature.
- To eliminate all extra cells and dye, rinse the plate three or four times by immersing it in a container of water as previously mentioned. Blot the plate well on a stack of paper towels after shaking off the water.
- After inverting the microtiter plate, it should be left to dry for a few hours or overnight.
- After the wells are totally dry, take pictures of them for qualitative examination.

3.3 Steps for Crystal Violet Staining Protocol

- For accurate and efficient staining, the crystal violet staining process often entails a number of crucial phases.
- These procedures include handling the post-staining procedure, prepping the cells, and applying the dye.
- Accurate findings in cell-based assays depend on each step being carried out carefully (*see* **Note 3**).

3.4 Cell Preparation

- A number of crucial steps are often involved in the crystal violet staining process in order to produce dependable results and efficient staining.
- These procedures involve getting the cells ready, putting the dye on them, and managing the post-staining procedure.
- In cell-based assays, meticulous execution of every step is essential to collecting reliable data.

3.5 Staining Procedure

- After preparing the cells, expose them to a 0.1% crystal violet solution.
- In the case of bacterial biofilms, gently add the solution to the wells containing the biofilm.
- Allow the staining process to proceed for 15–30 min at room temperature, during which the dye binds to cellular structures such as bacterial cell walls and the extracellular matrix within the biofilm.

3.6 Analyzing the Biofilm

1. To dissolve the bound crystal violet from the biofilm, 125 μL of a 30% acetic acid solution prepared in distilled water was carefully added to each well of the microtiter plate.
2. The addition of acetic acid facilitates the release of crystal violet from the stained cells or biofilm matrix, allowing it to go into solution for quantification.

3. After the reagent was added, the plate was left undisturbed at room temperature for 10–15 min to ensure complete solubilization of the dye.
4. Gentle incubation during this step is critical, as insufficient time may result in incomplete dye release, whereas overly long incubation may increase variability.
5. Once the dye was fully dissolved, 125 μL of the solubilized solution from each well was carefully transferred into a new flat-bottom microtiter plate.
6. The use of a fresh plate helps to avoid interference from residual biofilm material or uneven surfaces that might affect optical readings.
7. Absorbance was then measured at 550 nm using a plate reader, which provides a quantitative estimate of the amount of crystal violet retained, and thereby indirectly reflects the extent of biofilm biomass or cell density.
8. A blank control consisting of 30% acetic acid in water was included to calibrate the plate reader and eliminate background absorbance, ensuring that the readings accurately represented the dye bound to the cells or biofilm.

3.7 Post-Staining Handling

After staining, the processed cells must undergo a few essential steps to ensure reliable results.

1. Washing: Once incubation is complete, the staining solution is discarded, and the cells are washed four times with an appropriate washing solution. This step removes any unbound crystal violet, leaving only the dye that is firmly associated with the cells, which is crucial for accurate absorbance measurements.
2. Drying: The stained cells are then allowed to air dry. This step is particularly important in cell viability assays, as it prepares the samples for the subsequent solubilization phase.
3. Solubilization: To enable quantification, the retained dye is dissolved in a suitable solvent, such as ethanol or acetic acid. In the case of biofilms, an elution buffer like sodium citrate is often used to release the bound crystal violet effectively (*see* **Note 4**).

3.8 Biofilm Assay Protocol

1. Crystal violet staining is a widely used method for evaluating biofilm formation and involves several key steps.
2. Inoculation: A bacterial culture is added to the wells of a multi-well plate to enable biofilms to establish on the surface.
3. Biofilm development: The plate is incubated at 37 °C, either under static conditions or with gentle shaking, to encourage biofilm growth.

4. Staining: Once biofilms are formed, crystal violet is added to the wells and left to stain at room temperature for 15–30 min.
5. Washing: The wells are then gently rinsed to remove unattached or planktonic cells.
6. Quantification: The retained crystal violet is solubilized using ethanol or sodium citrate, and the absorbance is recorded at 590 nm to determine the extent of biofilm formation.

4 Notes

1. Sodium citrate is an optional reagent commonly used in biofilm assays. It helps elute crystal violet dye, enabling the quantification of biofilm formation through absorbance measurements.
2. Methanol and ethanol are also optional reagents, though their effectiveness differs. Methanol generally fixes cells more quickly, whereas ethanol provides a gentler fixation. These differences often determine which reagent is chosen based on the sample type and the intended results.
3. Certain wells appear darker, while some areas of the biofilm remain lightly stained. Make sure the crystal violet solution is applied uniformly to all wells, and maintain the same staining duration across samples. Using a multichannel pipette can help achieve even application.

Issue
Even mild rinsing can sometimes dislodge the biofilm, resulting in unreliable outcomes.

Remedy
Adjust the rinsing method by reducing the flow rate or minimizing the rinsing volume. Experiment with alternative rinsing solutions (such as PBS with different salt concentrations) to achieve the best compromise between eliminating excess stain and preserving biofilm integrity.

Issue
Natural variability in biofilm formation may occur across replicates.

Remedy
Perform several replicates for every condition and evaluate a greater number of wells. Incorporate a control group that uses a strain with established biofilm-forming ability.

4. Issue

Certain constituents of the growth medium may disrupt or affect crystal violet staining.

Remedy

Refine the washing process following biofilm growth but prior to staining to eliminate leftover medium. Alternatively, choose a growth medium less prone to affecting the staining outcome.

Issue

Errors may occur in precisely measuring the biofilm mass.

Remedy

Confirm that the crystal violet solution is fully dissolved and evenly mixed before measurement. Record absorbance with a spectrophotometer at the appropriate wavelength (typically 590–600 nm). To enhance precision, prepare a standard curve with defined crystal violet concentrations. Apply suitable statistical tools when analyzing the data. Through careful refinement of each stage in the crystal violet biofilm assay and by addressing these frequent challenges, more consistent and dependable results can be achieved.

Competing Interest

The author declares that there were no competing interests. All aspects of the work were conducted independently, without any financial, personal, or professional relationships that could be perceived as influencing the results, interpretation, or presentation of this study.

References

1. Costerton JW, Lewandowski Z, Caldwell DE, Korber DR, Lappin-Scott HM (1995) Microbial biofilms. Ann Rev Microbiol 49:711–745. https://doi.org/10.1146/annurev.mi.49.100195.003431
2. Jimenez-Fernandez A, Lopez-Sanchez A, Calero P, Govantes F (2015) The c-di-GMP phosphodiesterase BifA regulates biofilm development in pseudomonas putida. Environ Microbiol Rep 7:78–84
3. Christensen GD, Simpson WA, Younger JJ, Baddour LM, Barrett FF, Melton DM, Beachey EH (1985) Adherence of coagulase-negative Staphylococci to plastic tissue culture plates: a quantitative model for the adherence of Staphylococci to medical devices. J Clin Microbiol 22: 996–1006
4. O'Toole GA (2011) Microtiter dish biofilm formation assay. J Visual Exp (47):e2437. https://doi.org/10.3791/2437
5. Feoktistova M, Geserick P, Leverkus M (2016) Crystal violet assay for determining viability of cultured cells. Cold Spring Harbor Protocols 2016(4) pdb. prot087379. https://doi.org/10.1101/pdb.prot087379

Part III

Staining and Colorimetric Assays

Chapter 8

Fluorescence-Based Determination of Live and Dead Cells in Biofilm Communities

Vinushiya S, Priya P K, and Abirami K

Abstract

Biofilms represent organized microbial communities attached to surfaces and enclosed within a self-produced extracellular matrix. Cells growing in a biofilm differ significantly from their planktonic counterparts in terms of physiology, metabolic activity, and resistance to antimicrobial agents. Evaluating cell viability within these structured communities is essential for understanding biofilm development, persistence, and response to treatment.

Live/dead cell staining provides a rapid and reliable method to assess bacterial viability directly within intact biofilms. This technique is based on membrane integrity, where fluorescent nucleic acid dyes differentiate cells with intact membranes from those with compromised membranes. Following biofilm formation under defined conditions, samples are gently washed to remove non-adherent cells and then treated with a dual-fluorescent staining solution. After incubation in the dark, stained biofilms are visualized using fluorescence or confocal microscopy. Viable cells exhibit green fluorescence, whereas membrane-damaged or non-viable cells fluoresce red. The method preserves biofilm architecture and allows both qualitative observation and quantitative estimation of live and dead cell populations. It is particularly useful for studying antimicrobial efficacy, environmental stress responses, and biofilm structural organization. Overall, live/dead staining serves as a practical and informative approach for viability assessment in biofilm research.

Key words Biofilm, Cell viability, Live/dead staining, Fluorescence microscopy, Membrane integrity, Bacterial biofilm analysis, Confocal microscopy, Antimicrobial assessment

1 Introduction

Biofilms are structured microbial communities in which cells are irreversibly attached to a surface and embedded within a self-produced extracellular polymeric substance (EPS) matrix. Unlike free-floating (planktonic) cells, biofilm-associated microorganisms exhibit altered physiology, enhanced stress tolerance, and increased resistance to antimicrobial agents. Because of these properties, biofilms are of major interest in clinical, environmental, and industrial microbiology. Assessment of cell viability within a biofilm is

Abirami K (ed.), *Microbial Biofilm Research: Comprehensive Protocols and Best Practices*, Springer Protocols Handbooks, https://doi.org/10.1007/978-1-0716-5535-1_8,

essential for understanding microbial survival, antimicrobial efficacy, and biofilm dynamics. Conventional plate counting methods often underestimate viable cells in biofilms due to aggregation and the presence of viable but non-culturable (VBNC) populations. Fluorescence-based viability staining overcomes these limitations by allowing direct visualization of living and membrane-compromised cells within the intact biofilm structure [1].

The live/dead staining approach, commonly performed using the LIVE/DEAD BacLight Bacterial Viability Kit, is based on membrane integrity as an indicator of cell viability. The method employs two nucleic acid stains: SYTO 9, which penetrates all bacterial cells and emits green fluorescence, and propidium iodide, which enters only cells with damaged membranes and emits red fluorescence. When used together, these dyes provide a rapid and reliable means to distinguish viable from non-viable cells within biofilms. This technique is particularly valuable for biofilm studies because it enables in situ analysis without disrupting the three-dimensional architecture of the microbial community. When combined with fluorescence or confocal laser scanning microscopy, it allows both qualitative visualization and quantitative assessment of cell viability distribution throughout the biofilm matrix. Overall, live/dead staining is a straightforward and widely adopted method for evaluating biofilm viability under various experimental conditions, including antimicrobial treatment, environmental stress, and surface interaction studies [2–5].

2 Materials

- Biofilm samples (grown on coverslips, glass slides, microtiter plates, or other substrata).
- LIVE/DEAD BacLight staining kit.
- Sterile phosphate-buffered saline (PBS), pH 7.2–7.4.
- Sterile microcentrifuge tubes.
- Aluminum foil (for light protection).
- Glass slides and coverslips (if needed).
- Fluorescence or confocal laser scanning microscope.
- Pipettes and sterile tips.
- Buffer: Filter-sterilized 0.85 % NaCl or phosphate-buffered saline (PBS). Note: Some buffers like Citrate can interfere with staining; saline is generally preferred for biofilms.
- Microscopy: Confocal laser scanning microscope (CLSM) or an epifluorescence microscope.

3 Methods

Live/dead staining, formally known as Bacillary Viability Staining, is a fluorescence-based method used to differentiate between metabolically active (live) and membrane-compromised (dead) cells within a biofilm matrix. Unlike planktonic cells, biofilms present unique challenges due to the presence of extracellular polymeric substances (EPS), which can hinder dye penetration [6–8]. The differentiation relies on the physical integrity of the bacterial cell membrane (Fig. 1):

- *SYTO 9 (Green)*: This is a small, lipophilic molecule that can cross all bacterial cell membranes. Once inside, it binds to DNA and RNA, emitting a green signal. It labels all cells in the biofilm.
- *Propidium Iodide (Red)*: This is a large, charged molecule that is excluded by healthy, intact membranes. It can only enter cells with damaged membranes (dead or dying). Because PI has a higher affinity for nucleic acids than SYTO 9, it displaces the green dye when both are present in a cell [9, 10].

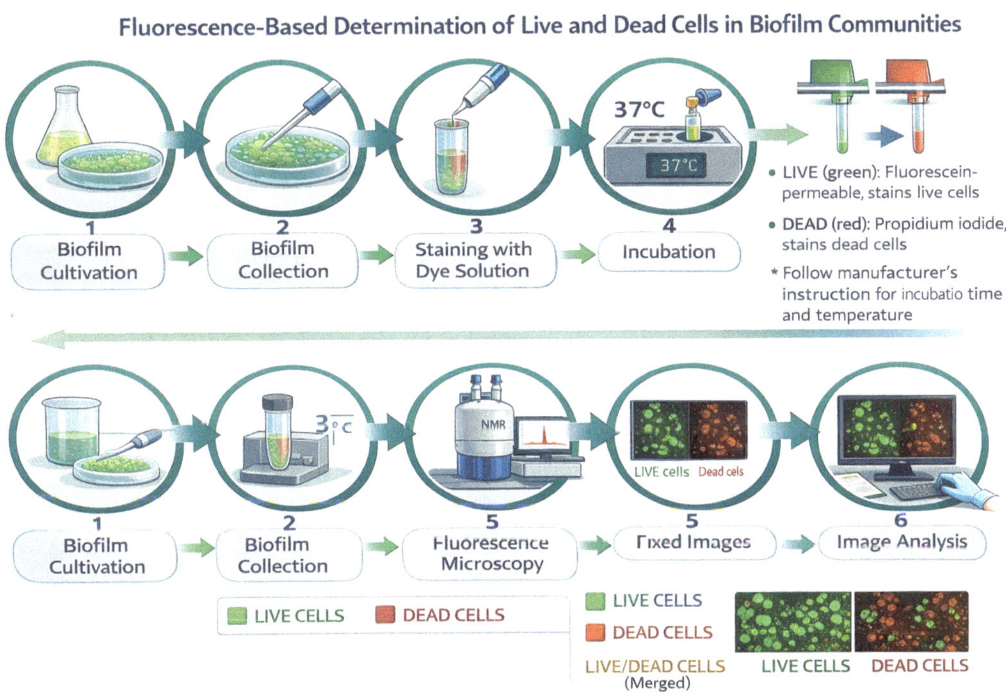

Fig. 1 Fluorescence-based determination of live and dead cells in biofilm communities. The schematic illustrates the stepwise protocol for assessing cell viability within biofilms using fluorescent staining and microscopy. Green fluorescence represents viable cells with intact membranes, whereas red fluorescence indicates membrane-compromised or dead cells. Merged images provide an overall assessment of biofilm viability and structural organization

Consequently, live cells appear green (≈500 nm), while dead cells appear red (≈635 nm).

3.1 Preparation of Biofilm

1. Grow biofilm under desired experimental conditions (e.g., 24–72 h incubation).
2. After incubation, gently remove planktonic cells by aspirating the culture medium.
3. Carefully rinse the biofilm 2–3 times with sterile PBS to remove loosely attached cells.
   - Avoid vigorous pipetting to prevent disruption of the biofilm structure (*see* **Note 1**)

3.2 Preparation of Staining Solution

1. Allow the dye components to equilibrate to room temperature.
2. Prepare a working staining solution by mixing (*see* **Note 2**):
   - 3 μL SYTO 9.
   - 3 μL Propidium iodide.
   - 1 mL sterile PBS.

   (Adjust volume proportionally depending on the number of samples.)
3. Mix gently by pipetting.
4. Protect the staining solution from light by covering with foil (*see* **Note 3**).

3.3 Staining Procedure

1. Add enough staining solution to fully cover the biofilm surface:
   - ~200 μL for a coverslip in a well plate.
   - ~100–150 μL for small surface areas.
2. Incubate the samples in the dark at room temperature for 15 min.
3. After incubation:
   - Gently remove excess stain.
   - Optionally rinse once with PBS to remove unbound dye (do not over-wash).
4. Mount the stained biofilm on a clean slide (if not already mounted).
   - Add a small drop of PBS to prevent drying.
   - Carefully place a coverslip on top.

3.4 Microscopy and Imaging

(a) Observe under a fluorescence microscope using:

- Green filter (approx. 480 nm excitation/500 nm emission) for live cells.
- Red filter (approx. 490–540 nm excitation/620–750 nm emission) for dead cells.

(b) For confocal microscopy:

- Adjust laser settings according to dye specifications.
- Capture Z-stack images if three-dimensional biofilm structure analysis is required.

(c) Acquire images from multiple random fields for quantitative analysis.

3.5 Data Interpretation

- Green fluorescence → Cells with intact membranes (viable).
- Red fluorescence → Cells with compromised membranes (non-viable).
- Mixed yellow/orange signal may indicate overlapping fluorescence or partially damaged cells.

Image analysis software (e.g., ImageJ) can be used to calculate live/dead ratios.

3.6 Controls (Recommended)

- Positive control (live cells): Fresh culture without treatment.
- Negative control (dead cells): Cells treated with 70% ethanol for 10–15 min before staining.
- Unstained control: To check background fluorescence.

4 Notes

1. *The "Ghost" Cell Problem*: In some biofilms, the EPS matrix contains extracellular DNA (eDNA). PI can bind to this eDNA, creating a red "cloud" around the cells that might be misinterpreted as a high death rate.
2. *Dye Penetration*: For very thick biofilms (>100 μm), the dyes may not reach the innermost layers. If the center of a thick cluster appears dark, it is likely a penetration issue rather than a lack of cells.
3. *Intermediate States*: You may observe yellow or orange cells. These are often "injured" cells where the membrane is partially compromised, allowing some PI to enter but not enough to fully displace the SYTO 9.
4. *Autofluorescence*: Some components of the EPS or certain bacterial species (like *Pseudomonas*) naturally fluoresce. Always run an unstained control to check for background signal.

Competing Interest

The authors report no financial conflicts related to the subject matter of this chapter.

References

1. Thermo Fisher Scientific (2004) LIVE/DEAD® BacLight™ Bacterial Viability Kits Product Information. Molecular Probes, Inc. [Revised Manual]
2. Lawrence JR, Neu TR (1999) Confocal laser scanning microscopy for analysis of microbial biofilms. Methods Enzymol 310:131–144. https://doi.org/10.1016/S0076-6879(99)10011-8
3. Boulos L, Prévost M, Barbeau B, Coallier J, Desjardins R (1999) LIVE/DEAD BacLight: application of a new rapid staining method for direct enumeration of viable and total bacteria in drinking water. J Microbiol Methods 37(1): 77–86. https://doi.org/10.1016/s0167-7012(99)00048-2
4. Netuschil L, Reich E, Unteregger G, Sculean A, Brecx M (1998) A pilot study of confocal laser scanning microscopy for the assessment of undisturbed dental plaque vitality and topography. Arch Oral Biol 43(4): 277–285. https://doi.org/10.1016/s0003-9969(97)00121-0
5. Whitchurch CB, Tolker-Nielsen T, Ragas PC, Mattick JS (2002) Extracellular DNA required for bacterial biofilm formation. Science 295(5559):1487. https://doi.org/10.1126/science.295.5559.1487
6. Berney M, Hammes F, Bosshard F, Weilenmann HU, Egli T (2007) Assessment and interpretation of bacterial viability by using the LIVE/DEAD BacLight kit in combination with flow cytometry. Appl Environ Microbiol 73(10):3283–3290. https://doi.org/10.1128/aem.02750-06
7. Donlan RM (2002) Biofilms: microbial life on surfaces. Emerg Infect Dis 8(9):881–890. https://doi.org/10.3201/eid0809.020063
8. O'Toole GA (2011) Microtiter dish biofilm formation assay. J Vis Exp (47):2437. https://doi.org/10.3791/2437
9. Walters MC 3rd, Roe F, Bugnicourt A, Franklin MJ, Stewart PS (2003) Contributions of antibiotic penetration, oxygen limitation, and low metabolic activity to tolerance of Pseudomonas aeruginosa biofilms to ciprofloxacin and tobramycin. Antimicrob Agents Chemother 47(1):317–323. https://doi.org/10.1128/aac.47.1.317-323.2003
10. Walters MCRF, Bugnicourt A, Franklin MJ, Stewart PS (2003) Contributions of antibiotic penetration, oxygen limitation, and low metabolic activity to tolerance of *Pseudomonas aeruginosa* biofilms to ciprofloxacin and tobramycin. Antimicrob Agents Chemother 47:317–323. https://doi.org/10.1128/aac.47.1.317-323.2003

Chapter 9

Rapid Colorimetric Methods for Evaluating Biofilm Viability Using Tetrazolium and Resazurin Dyes

Sujith P

Abstract

Rapid and reliable evaluation of microbial viability in biofilms is essential for antimicrobial screening, probiotic assessment, and environmental microbiology studies. Tetrazolium- and resazurin-based colorimetric assays, including XTT, MTT, and Resazurin methods, provide sensitive and high-throughput approaches to quantify metabolic activity in both planktonic and biofilm-forming microorganisms such as *Priestia megaterium*. These assays rely on the reduction of redox-sensitive dyes by cellular respiratory enzymes, producing measurable color changes proportional to viable cell activity. Tetrazolium salts are converted into colored formazan products, while resazurin is reduced to fluorescent resorufin, enabling rapid spectrophotometric or fluorometric analysis. Compared with conventional biomass staining techniques, these methods are faster, cost-effective, and suitable for microtiter plate screening, making them valuable tools for studying antimicrobial efficacy, biofilm resistance, stress responses, and microbial physiology. When combined with complementary biomass quantification methods, tetrazolium and resazurin assays provide a comprehensive understanding of biofilm viability and metabolic function, supporting their widespread application in clinical, industrial, and environmental biofilm research.

Key words XTT assay, MTT assay, Resazurin assay, Tetrazolium reduction, Biofilm metabolic activity

1 Introduction

Tetrazolium and resazurin based colorimetric assays are widely used to evaluate microbial viability and metabolic activity in biofilms and planktonic cultures. These methods rely on the reduction of redox-sensitive dyes by cellular respiratory enzymes, producing measurable color changes proportional to metabolic activity [1]. In the XTT and MTT assays, tetrazolium salts are reduced by NADH-dependent electron transport systems to colored formazan products, whereas in the resazurin assay, viable cells convert blue resazurin into pink resorufin through intracellular redox reactions [2]. Because these assays measure metabolic activity rather than total biomass, they are particularly useful for assessing antimicrobial

Abirami K (ed.), *Microbial Biofilm Research: Comprehensive Protocols and Best Practices*, Springer Protocols Handbooks, https://doi.org/10.1007/978-1-0716-5535-1_9,

efficacy, probiotic performance, and stress responses in biofilm-forming microorganisms such as *Priestia megaterium*.

Compared with traditional staining techniques, XTT, MTT, and resazurin assays are rapid, sensitive, and suitable for high-throughput screening using microtiter plates. They are therefore widely applied in studies of antimicrobial resistance, antibiofilm compound screening, medical device contamination, and environmental microbiology [3]. Combining these metabolic assays with complementary biomass measurements provides a comprehensive understanding of biofilm viability and functional activity, making them essential tools in modern biofilm research.

2 Materials

2.1 Reagents

- XTT sodium salt
- MTT (3-(4,5-dimethylthiazol-2-yl)-2,5-diphenyltetrazolium bromide)
- Resazurin sodium salt (Alamar Blue reagent)
- Menadione (electron mediator for XTT assay)
- Dimethyl sulfoxide (DMSO) or Isopropanol (for MTT formazan solubilization)
- Phosphate Buffered Saline (PBS), pH 7.2
- Nutrient Broth/Tryptic Soy Broth/MRS Broth (depending on organism)
- Sterile distilled water
- Antimicrobial test compounds (if evaluating inhibition)
- Standard biofilm-forming culture such as *Priestia megaterium*

2.2 Equipment

- 96-well sterile flat-bottom microtiter plates
- Microplate reader (490 nm for XTT, 570 nm for MTT, 570–600 nm for resazurin)
- Incubator (30–37 °C)
- Micropipettes (10–1000 μL)
- Sterile pipette tips
- Vortex mixer
- Centrifuge (optional)
- Aluminum foil (to protect reagents from light)

2.3 Preparation of Stock Solutions

XTT Stock Solution

- Dissolve XTT in PBS to 1 mg/mL
- Filter sterilizes and store at −20 °C in dark

Menadione Stock

- Prepare 10 mM menadione in acetone
- Store at −20 °C, light-protected

MTT Stock Solution

- Dissolve MTT in PBS to 5 mg/mL
- Filter sterilizes; store at 4 °C in dark

Resazurin Solution

- Prepare 0.01% resazurin in sterile distilled water
- Store at 4 °C, protected from light

3 Methods

Tetrazolium salts such as XTT and MTT are reduced by NADH-dependent cellular respiration to colored formazan products, providing an indirect measure of metabolically active cells. In the resazurin assay, viable cells reduce blue resazurin to pink resorufin. These assays quantify metabolic activity in biofilms and planktonic cultures and are widely used for antimicrobial screening and biofilm studies [4, 5].

3.1 Biofilm Formation

1. Inoculate a single colony of the test organism (e.g., *Priestia megaterium*) into 5–10 mL of appropriate sterile broth medium and incubate overnight (16–18 h) at 37 °C under shaking conditions (150–200 rpm) to obtain a fresh, actively growing culture.
2. Measure the optical density of the overnight culture at 600 nm using a spectrophotometer. Dilute the culture with sterile fresh broth until an OD_{600} of 0.5 is achieved, ensuring a standardized bacterial concentration for consistent biofilm formation. Add 200 μL suspension into sterile 96-well plate.
3. Dispense 200 μL of the adjusted bacterial suspension into each well of a sterile, flat-bottom 96-well polystyrene microtiter plate. Include appropriate controls such as uninoculated medium (negative control) and, if required, treatment wells. Remove planktonic cells and wash wells gently 2–3× with PBS.
4. Incubate the plate statically at 37 °C for 24–48 h without shaking to allow bacterial adhesion and biofilm development on the well surface. Avoid disturbing the plate during incubation to ensure uniform biofilm formation.
5. After incubation, carefully aspirate or decant the planktonic (non-adherent) cells without disturbing the attached biofilm. Gently wash the wells 2–3 times with sterile phosphate-

Table 1
Comparison of colorimetric assays (MTT, XTT and Resazurin Assays)

Feature	MTT assay	XTT assay	Resazurin assay
Principle	Reduction of yellow MTT to insoluble purple formazan crystals by metabolically active cells	Reduction of XTT to a soluble orange-colored formazan product	Reduction of blue resazurin to pink fluorescent resorufin by viable cells
Formazan solubility	Insoluble (requires solubilization step)	Soluble (no additional solubilization required)	Soluble end product
Additional reagent required	No electron mediator required	Requires electron-coupling agent (e.g., menadione)	No additional mediator required
Detection method	Absorbance at ~570 nm	Absorbance at ~490 nm	Absorbance (~570 nm) or fluorescence (Ex 560 nm/Em 590 nm)
Sensitivity	Moderate	Higher than MTT	High (especially in fluorescence mode)
Processing time	3–4 h incubation + solubilization time	2–3 h incubation	1–2 h incubation
Ease of procedure	More steps due to crystal dissolution	Simpler than MTT	Simplest and rapid procedure
Suitability for biofilms	Suitable but may underestimate activity due to diffusion limits	Well suited for biofilms due to soluble product	Highly suitable; good penetration and sensitive detection
Advantages	Widely established, reproducible	No crystal dissolution step; more convenient	Non-destructive, rapid, highly sensitive, fluorescence option
Limitations	Requires dissolution step; possible incomplete solubilization	Requires fresh mediator; light-sensitive reagents	Dye reduction rate may vary with cell density and conditions

Summary of the underlying principles, detection methods, sensitivity, procedural complexity, advantages, and limitations of three commonly used colorimetric and fluorometric assays (MTT, XTT, and resazurin) employed to evaluate biofilm viability and metabolic activity in microtiter plate-based experiments

buffered saline (PBS) to remove loosely attached bacteria. Take care to avoid touching the well bottom to prevent disruption of the biofilm layer (Table 1).

3.2 XTT Assay

1. Dissolve XTT sodium salt in sterile phosphate-buffered saline (PBS) to obtain a final concentration of 1 mg/mL. Filter-sterilize the solution using a 0.22 μm syringe filter to remove particulates and ensure sterility. Prepare fresh before use or store aliquots at −20 °C protected from light.

2. Prepare a fresh menadione stock solution (e.g., 10 mM) in acetone or dimethyl sulfoxide (DMSO). Immediately before use, add menadione to the XTT solution to achieve a final concentration of 10 μM. Mix gently to ensure uniform distribution. This working solution should be prepared fresh and protected from light, as both XTT and menadione are light-sensitive.
3. After washing the pre-formed biofilms to remove planktonic cells, add 200 μL of the freshly prepared XTT–menadione solution to each well of the 96-well plate. Ensure that the biofilm surface is completely covered by the reagent.
4. Incubate the plate at 37 °C for 2–3 h in the dark under static conditions. During incubation, metabolically active cells within the biofilm reduce XTT to a soluble orange-colored formazan product. Avoid disturbing the plate to prevent disruption of the biofilm.
5. After incubation, carefully pipette 150 μL of the colored supernatant from each well into a new sterile 96-well plate. Avoid transferring any biofilm material to ensure accurate absorbance readings.
6. Measure the absorbance of the transferred supernatant at 490 nm using a microplate reader. The absorbance values correlate with the metabolic activity of the biofilm. Include appropriate blanks (XTT–menadione solution without cells) and controls for background correction (*see* **Note 3**).

3.3 MTT Assay

1. A fresh MTT solution (0.5 mg/mL) was prepared in sterile PBS or suitable medium. After washing pre-formed biofilms to remove non-adherent cells, 200 μL of MTT solution was added to each well.
2. The plates were incubated at 37 °C for 3–4 h under static conditions to allow viable cells to convert MTT into insoluble purple formazan crystals.
3. After incubation, the MTT solution was carefully removed without disturbing the crystal deposits. To dissolve the formazan crystals, 200 μL of dimethyl sulfoxide (DMSO) or isopropanol was added to each well.
4. Plates were gently agitated to ensure complete solubilization and uniform color distribution.
5. Absorbance was recorded at 570 nm using a microplate reader. The measured optical density was directly proportional to the metabolic activity and viability of the biofilm cells (*see* **Note 3**).

3.4 Resazurin Assay

1. Prepare a fresh 0.01% (w/v) resazurin solution in sterile phosphate-buffered saline (PBS) or appropriate culture medium. After removing planktonic cells and gently washing the established biofilms with PBS to eliminate non-adherent bacteria, dispense 200 μL of the resazurin solution into each well of the 96-well microtiter plate. Ensure that the entire biofilm surface is fully covered by the reagent (*see* **Note 2**).
2. Incubate the plate at 37 °C for 1–2 h under static conditions. During incubation, metabolically active cells within the biofilm reduce the blue, non-fluorescent resazurin dye to pink, fluorescent resorufin. The extent of color change is proportional to the cellular metabolic activity. To prevent variability, protect the plate from direct light exposure during incubation (*see* **Note 1**).
3. Following incubation, assess the conversion of resazurin by measuring either absorbance at 570 nm (with a reference wavelength if required) or fluorescence using excitation at 560 nm and emission at 590 nm. Fluorescence detection generally provides higher sensitivity. Include reagent blanks (resazurin without cells) for background correction. The resulting values reflect the metabolic activity and viability of the biofilm cells (*see* **Note 4**) (Fig. 1).

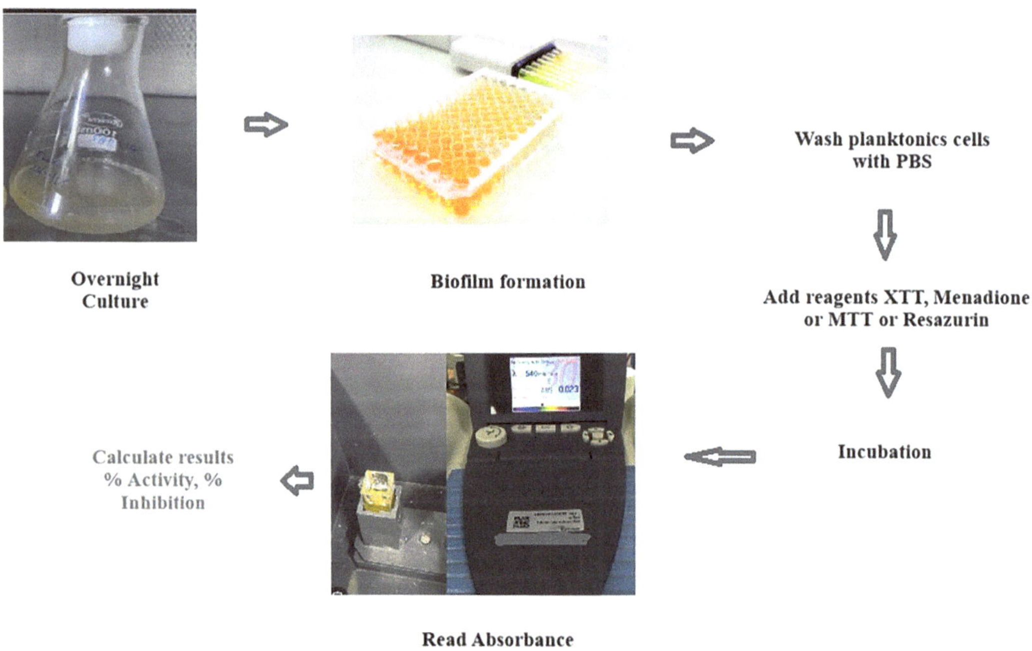

Fig. 1 Laboratory workflow for XTT, MTT, and resazurin assays used to quantify metabolic activity in biofilm-forming microorganisms grown in microtiter plates

Calculations

Relative Metabolic Activity (%)

$$\text{Activity} = \frac{OD_{sample}/OD_{treated} \times 100}{OD_{Control}}$$

Where

$OD_{treated}$ = Optical density of treated sample
$OD_{Control}$ = Optical density of untreated control

Biofilm Inhibition (%)

$$\text{Inhibition} = \frac{OD_{control} - OD_{treated} \times 100}{OD_{Control}}$$

Where

$OD_{Control}$ = Optical density of control biofilm
$OD_{treated}$ = Optical density after treatment

For example

- OD of Control = 0.80
- OD of treated sample = 0.32

Relative Metabolic Activity (%)

$$\text{Activity} = \frac{OD_{sample}/OD_{treated} \times 100}{OD_{Control}}$$
$$= 0.32/0.80 \times 100$$
$$= 0.4 \times 100$$

Relative Metabolic Activity (%) = 40%

Biofilm Inhibition (%)

$$\text{Inhibition} = \frac{OD_{control} - OD_{treated} \times 100}{OD_{Control}}$$
$$= \frac{0.80 - 0.32}{0.80}$$
$$= 0.48/0.80 \times 100$$
$$= 0.6 \times 100$$

Biofilm inhibition = 60%

4 Notes

1. Tetrazolium salts such as MTT and XTT are sensitive to light exposure, which can lead to premature degradation and reduced assay performance. To maintain reagent stability and ensure reliable results, stock and working solutions should be prepared under minimal light conditions and stored in amber containers or wrapped in aluminum foil. During incubation, plates should also be protected from direct light.
2. Prior to adding metabolic reagents, wells should be carefully rinsed with sterile phosphate-buffered saline (PBS) to remove residual growth medium, planktonic cells, and any treatment compounds that may interfere with dye reduction or absorbance readings. Gentle washing is essential to prevent disruption of the attached biofilm while ensuring accurate measurement of metabolic activity.
3. Each experimental condition should be tested in at least three technical replicates within the same plate to minimize variability and improve data reliability. In addition, conducting independent biological repeats strengthens reproducibility and allows for appropriate statistical analysis of the results.
4. Since metabolic assays measure cellular activity rather than total biofilm mass, it is advisable to complement them with a biomass-based staining method, such as crystal violet staining. This combined approach provides a more comprehensive evaluation by distinguishing between changes in biofilm viability and alterations in total biofilm accumulation.

Competing Interest

The author declares that there are no financial conflicts of interest.

References

1. Riss TL, Moravec RA, Niles AL et al (2016) Cell viability assays. In: Sittampalam GS, Coussens NP, Brimacombe K et al (eds) Assay guidance manual. Eli Lilly & Company and the National Center for Advancing Translational Sciences, Bethesda
2. Rampersad SN (2012) Multiple applications of Alamar Blue as an indicator of metabolic function and cellular health in cell viability bioassays. Sensors 12(9):12347–12360. https://doi.org/10.3390/s120912347
3. O'Toole GA (2011) Microtiter dish biofilm formation assay. J Vis Exp 47:e2437. https://doi.org/10.3791/2437
4. Merritt JH, Kadouri DE, O'Toole GA (2005) Growing and analyzing static biofilms. Curr Protoc Microbiol Chapter 1(Unit 1B.1). https://doi.org/10.1002/9780471729259.mc01b01s00
5. O'Brien J, Wilson I, Orton T, Pognan F (2000) Investigation of the alamar blue (resazurin) fluorescent dye for the assessment of mammalian cell cytotoxicity. Eur J Biochem 267:5421–5426. https://doi.org/10.1046/j.1432-1327.2000.01606.x

Part IV

Molecular and Genetic Methods

Chapter 10

Polymerase Chain Reaction at the Biofilm Interface: Tools for Detection, Quantification, and Analysis

Venkatesh Kumaresan

Abstract

Microbial biofilms are highly structured, surface-associated communities characterized by spatial heterogeneity, metabolic diversity, and enhanced tolerance to antimicrobial stress. Comprehensive biofilm research, therefore, requires molecular tools that enable not only sensitive detection and accurate quantification but also functional and compositional characterization of biofilm-associated microorganisms. Polymerase chain reaction (PCR)-based approaches have become central to biofilm analysis due to their specificity, scalability, and compatibility with complex biofilm matrices. This chapter provides a detailed methodological framework for applying conventional endpoint PCR, quantitative real-time PCR (qPCR), and droplet digital PCR (ddPCR) to the detection, quantification, and molecular characterization of microbial biofilms. The topics covered include primer and probe design, biofilm-specific sample preparation, reaction setup, thermal cycling parameters, data analysis strategies, quality control, and troubleshooting. Practical examples demonstrate PCR-based qualitative assessment of biofilm-related genes in *Staphylococcus aureus* and qPCR-based quantitative assessment of biofilm-related gene expression in *Pseudomonas aeruginosa*, highlighting normalization strategies and data interpretation. This chapter also discusses selection of qPCR chemistries and outlines the advantages of ddPCR for absolute quantification and precise characterization of low-abundance targets. Overall, this chapter serves as a practical reference for researchers investigating biofilm structure, dynamics, and functional activity using PCR-based molecular tools. Here, we describe PCR-based methods for the detection, quantification, and molecular characterization of microbial biofilms.

Key words Digital PCR (dPCR), Biofilm interface, Bacterial load quantification, Biofilm matrix disruption, Standard curve analysis

1 Introduction

1.1 Background

Microbial biofilms are structured communities of microorganisms encased in a self-produced extracellular polymeric substance (EPS) and attached to biotic or abiotic surfaces [1, 2]. These biofilms are ubiquitous in natural, industrial, and clinical settings and contribute to a variety of processes ranging from nutrient cycling in ecosystems to chronic infections in humans. Biofilm-associated microbes, including bacteria, algae, and fungi exhibit enhanced

Abirami K (ed.), *Microbial Biofilm Research: Comprehensive Protocols and Best Practices*, Springer Protocols Handbooks,
https://doi.org/10.1007/978-1-0716-5535-1_10,

resistance to drugs and host immune responses, making them challenging to eradicate and a major concern in both healthcare and environmental microbiology [3, 4]. In aquatic organisms, the skin functions as a dynamic mucosal barrier that restricts microbial colonization and biofilm maturation by pathogenic bacteria [5]; however, several biofilm-forming microbes are able to overcome this defense and pose a significant threat [6].

1.2 Why PCR/qPCR for Biofilm Analysis?

PCR and qPCR are particularly valuable in biofilm research for several reasons:

1. *Detection and Identification:* PCR, especially for rRNA-targeted sequencing allows for the identification of microbial species within complex biofilm communities, even when traditional culturing methods fail [7].
2. *Quantitative Analysis:* qPCR facilitates accurate quantification of microbial load and gene expression levels within biofilms, including comparisons between planktonic and sessile states [8].
3. *Gene Expression Profiling:* qPCR is used to measure transcriptional activity of genes involved in biofilm formation (e.g., quorum sensing, EPS production, stress response) [9].
4. *Antimicrobial Resistance Monitoring:* Biofilms frequently harbor multidrug-resistant organisms. PCR-based methods enable detection of resistance genes directly from biofilm matrices [10].
5. *Clinical and Environmental Relevance:* These molecular tools can be applied to biofilms on medical devices, water pipelines, soil surfaces, and industrial equipment, making the protocols broadly applicable.

However, biofilm samples present unique technical challenges. The dense EPS matrix can interfere with nucleic acid extraction, and inhibitors within the biofilm may affect PCR efficiency [11]. Therefore, optimizing protocols specifically for biofilm-derived samples is critical for accurate and reproducible results.

This chapter outlines standardized and customizable protocols for PCR and qPCR analyses of microbial biofilms, with guidance on sample preparation, nucleic acid extraction, primer design, and data interpretation. Special attention is given to the technical hurdles specific to biofilm matrices and strategies to overcome them.

2 Materials and Reagents

This section lists the essential reagents, consumables, and equipment required for molecular analysis of microbial biofilms. Due to the complex structure of biofilms and the presence of inhibitory substances in the extracellular matrix, materials should be selected to maximize nucleic acid purity and amplification efficiency.

2.1 Reagents

A. Common Reagents for Biofilm Disruption and Nucleic Acid Extraction

These are used for both PCR and qPCR workflows.

- *Phosphate-buffered saline (PBS), sterile*—for biofilm washing.
  - To make 1 L of 10X PBS stock solution, combine 17.8 g of Na_2HPO_4, 2.4 g of KH_2PO_4, 80 g of NaCl, 2 g of KCl, and adjust final volume to 1 L.
  - To make 1 L of 1X PBS, add 100 mL of 10X PBS to 900 mL of water.
- *Lysozyme (10–50 mg/mL)*—enzymatic digestion of Gram-positive cell walls.
- *Proteinase K (20 mg/mL)*—protein digestion.
- *EDTA, 10–100 mM*—chelates divalent ions, may help disrupt EPS.
- *Sodium meta-periodate or dispersin B* (optional)—EPS disruption [12].
- *Bead-beating matrix tubes*—for mechanical lysis of tough biofilms.
- *DNA extraction kits:*
  - *DNeasy PowerBiofilm Kit* (Qiagen).
  - *ZymoBIOMICS DNA Miniprep Kit.*
- *RNA extraction kits* (for qPCR applications):
  - *RNeasy PowerBiofilm Kit* (Qiagen).
  - *TRIzol™ Reagent* (Thermo Fisher).
- *DNase I (RNase-free)*—to remove DNA from RNA samples.
- *RNase A*—to remove RNA from DNA samples.
- *Ethanol (70% and 100%), isopropanol, nuclease-free water.*

B. Reagents Specific to PCR

Used for endpoint (conventional) PCR to amplify and visualize DNA fragments.

- *Taq DNA polymerase* (standard or hot-start).
- *High-fidelity polymerase* (e.g., Phusion, Q5)—for cloning or sequencing applications.
- *10× PCR buffer with $MgCl_2$.*
- *dNTP mix (10 mM each).*
- *Gene-specific primers* (forward and reverse, 10 μM working stocks).

- *Nuclease-free water.*
- *Gel electrophoresis reagents:*
  - Agarose.
  - Ethidium bromide (5 mg/mL) or SYBR Safe DNA gel stain.
  - 1× TAE or TBE buffer.
  - DNA ladder (100 bp or 1 kb).

C. Reagents Specific to qPCR

Used for quantitative gene expression or copy number analysis.

- *Reverse transcription kit* (for cDNA synthesis from RNA):
  - *SuperScript IV* (Invitrogen).
  - *iScript*™ *cDNA Synthesis Kit* (Bio-Rad).
- *qPCR master mixes:*
  - *SYBR Green-based:*
    - *PowerUp SYBR Green Master Mix* (Thermo Fisher).
    - *QuantiTect SYBR Green Kit* (Qiagen).
  - *Probe-based:*
    - *TaqMan Universal PCR Master Mix* (Thermo Fisher).
- *qPCR primers or probes.*
- *ROX reference dye* (if required by the qPCR platform).
- *Standards or serial dilutions* for quantification.
- *Housekeeping gene primers* for normalization (e.g., *rpoB*, *gyrB*, *16S rRNA*) [13].

D. Reagents Specific to ddPCR

Most materials overlap with qPCR but require ddPCR-compatible formulations:

- ddPCR Supermix (e.g., *Bio-Rad QX200 ddPCR EvaGreen or Probe Supermix*).
- DNA or cDNA template.
- Gene-specific primers or probes (same design rules as qPCR).
- Nuclease-free water.

2.2 Consumables

- Sterile microcentrifuge tubes (1.5 mL, 2.0 mL).
- PCR tubes (0.2 mL) or qPCR 96-well plates.
- Optical adhesive film (for qPCR plates).
- Filtered, nuclease-free pipette tips.
- Sterile scrapers or pipette tips for biofilm collection.

- Gloves, lab coats, and barrier tips to avoid contamination.
- ddPCR cartridges and gaskets.
- Droplet generation oil.
- ddPCR 96-well plates.
- Plate seals.

2.3 Equipment

- *Common Equipment:*
  - Microcentrifuge.
  - Vortex mixer.
  - Biosafety cabinet.
  - Bead beater or homogenizer.
  - Nanodrop or Qubit fluorometer.
- *PCR-Specific Equipment:*
  - Thermocycler (conventional PCR).
  - Gel electrophoresis apparatus and power supply.
  - UV transilluminator or gel documentation system.
- *qPCR-Specific Equipment:*
  - Real-time PCR machine (with appropriate detection channels for SYBR or TaqMan assays).
- *ddPCR-Specific Equipment,*
  - Droplet generator.
  - PCR thermocycler.
  - Droplet reader.
  - Analysis software (e.g., *QuantaSoft or ddPCR Analysis Pro*).

3 PCR Method for Microbial Biofilms

Polymerase chain reaction (PCR) is a powerful method for detecting specific microbial DNA within biofilms. Given the structural complexity of biofilms and their potential to contain PCR inhibitors, optimizing both sample preparation and reaction setup is essential for successful amplification (Fig. 1).

3.1 Primer Design Criteria

Effective primer design is critical for specificity, efficiency, and reproducibility of PCR. When targeting genes in biofilm-associated microbes, consider the following:

3.1.1 Target Gene Selection

- Choose conserved genes for species identification (e.g., *16S rRNA, gyrB, rpoB*).

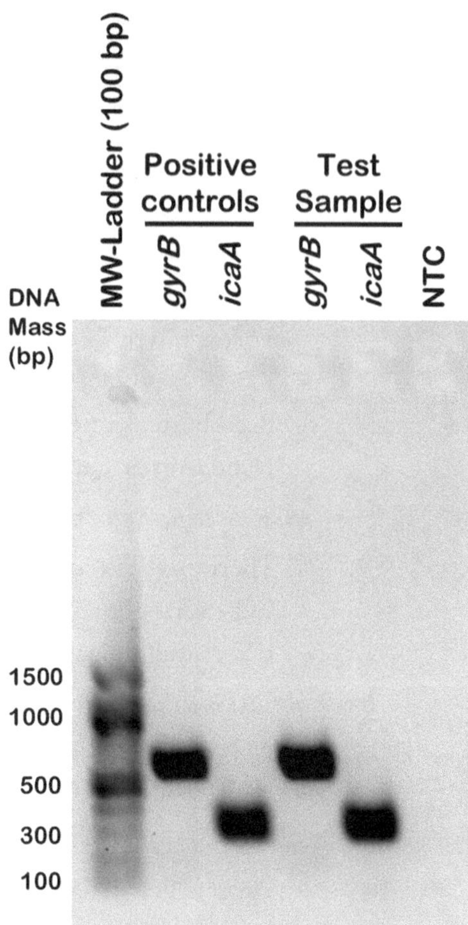

Fig. 1 PCR amplification of a biofilm-associated gene and a reference gene in biofilm-forming bacteria. PCR products resolved on a 1% agarose gel corresponding to the biofilm-associated gene *icaA* (300 bp) and the reference gene *gyrB* (600 bp) from a clinical isolate of *Staphylococcus aureus* amplified using gene-specific primers are shown. Lane 1: 100 bp molecular weight ladder; lanes 2 and 3: *gyrB* and *icaA*, respectively, in a known sample (positive controls); lanes 4 and 5: *gyrB* and *icaA*, respectively, in test sample. Lane 6: NTC

- Use biofilm-associated genes (e.g., *icaA, algD, lasR, pelA*) for functional studies.
- Validate gene presence in the microbial genome(s) of interest using NCBI BLAST or databases like IMG or KEGG.

3.1.2 Primer Design Guidelines

- *Length:* 18–25 nucleotides.
- *GC Content:* 40–60%.
- *Melting Temperature (Tm):* 55–65 °C (primers should not differ by more than 2 °C).

- Online Tm calculator can be used (e.g., https://tmcalculator.neb.com/).

- *Avoid:*
  - Runs of 4+ identical nucleotides.
  - Secondary structures (hairpins, self-dimers, cross-dimers).
  - Complementarity at the 3′ ends (to prevent primer-dimer formation).
- *Amplicon Size:*
  - 100–500 bp for most detection assays.
  - <300 bp preferred for degraded or environmental samples.
- Use primer design tools:
  - *Primer3, NCBI Primer-BLAST, or OligoAnalyzer* (IDT).

3.2 Sample Preparation for PCR

1. *Biofilm Collection* [14]:
   - Wash biofilm-grown surfaces with PBS to remove planktonic cells.
   - Scrape biofilm using sterile scalpel or pipette tip into 1.5 mL microcentrifuge tubes.
2. *Cell Disruption:*
   - Use enzymatic digestion (e.g., lysozyme + Proteinase K) and/or bead-beating.
3. *DNA Extraction:*
   - Follow a commercial kit protocol or phenol: chloroform method.
   - Include a final cleanup step to remove inhibitors (e.g., ethanol precipitation or column purification) (*see* **Note 1**).
4. *DNA Quantification:*
   - Measure DNA concentration using Nanodrop or Qubit.
   - Normalize all samples to equal DNA input (e.g., 10–50 ng per reaction).

3.3 PCR Reaction Setup

Prepare the PCR master mix on ice as follows (example for 25 μL reaction volume):

Component	Volume
Nuclease-free water	to 25 μL
10× PCR buffer	2.5 μL

(continued)

Component	Volume
dNTP mix (10 mM each)	0.5 μL
Forward primer (10 μM)	0.5 μL
Reverse primer (10 μM)	0.5 μL
Taq DNA polymerase (1–2.5 U)	0.25 μL
$MgCl_2$ (25 mM, if separate)	1.5 μL (final 1.5–2.5 mM)
Template DNA	1–2 μL (10–50 ng)

Mix gently and centrifuge briefly.

3.4 Thermocycling Conditions

Step	Temperature	Time	Cycles
Initial denaturation	95 °C	2–5 min	1
Denaturation	94–95 °C	30 sec	30–35
Annealing	50–65 °C[a]	30 sec	
Extension	72 °C	30–60 sec per kb	
Final extension	72 °C	5–10 min	1
Hold	4 °C	∞	

[a]Choose annealing temperature based on primer Tm (typically Tm—2 to 5 °C). Perform gradient PCR if optimization is needed

3.5 PCR Product Analysis

1. Prepare a 1–2% agarose gel with a suitable DNA stain (e.g., SYBR Safe or ethidium bromide).
2. Load 5–10 μL of PCR product with loading dye.
3. Run gel at 90–120 V for 30–60 min.
4. Visualize bands using a gel documentation system.
5. Confirm expected amplicon size using a DNA ladder (*see* **Note 2**).

3.6 Controls

- *Negative Control (NTC):* Reaction without DNA template to monitor contamination.
- *Positive Control:* DNA from a known positive strain/sample.
- *Extraction Control (optional):* Include a sample where no biofilm was processed to assess contamination in extraction reagents.

3.7 Nested PCR

Nested PCR is employed to enhance the sensitivity and specificity of detecting low-abundance bacterial DNA in biofilm samples, particularly when microorganisms are embedded within an extracellular polymeric substance (EPS) matrix [15]. Failure to obtain

Table 1
Summary comparison table

Metric	PCR	qPCR	ddPCR
Output	Gel band (size-based)	Ct value, amplification curve	Absolute copy number
Quantification	Semi-quantitative	Relative or absolute (with standard curve)	Absolute (Poisson-based)
Sensitivity	Moderate	High	Very high
Specificity	Primer-dependent	Primer ± probe-dependent	Highest (probe + droplet partitioning)
Ideal use	Gene presence, screening, cloning	Expression profiling, relative quant	Rare targets, low expression, inhibitor-tolerant quant

amplification in conventional PCR may result from insufficient genomic DNA template, a common limitation when analyzing biofilm-associated bacteria. In such cases, nested PCR provides an effective solution by amplifying target sequences in two sequential rounds. In the first round, a longer DNA fragment is amplified using outer primers targeting a conserved bacterial gene. The resulting amplicon then serves as the template for a second round of PCR using internal (nested) primers designed to amplify a shorter, more specific fragment (Table 1).

- *First-Round PCR:*

  The initial PCR is performed following standard PCR conditions using outer primers amplifying a bigger fragment (e.g., 500 bp) of the target bacterial gene.
- *Second-Round (Nested) PCR:*

  An aliquot (1–2 μL) of the first-round PCR product is used as the template for the nested PCR with internal primers amplifying a smaller fragment (e.g., 200 bp) from the product of first-round PCR. Cycling conditions are similar to the first round but typically use fewer cycles (25–30) to minimize nonspecific amplification.

- *Product Analysis and Controls:*

  Nested PCR products are analyzed by agarose gel electrophoresis. Positive controls, negative controls, and no-template controls are included in both PCR rounds to ensure specificity and to monitor potential contamination (*see* **Note 3**).

3.8 Quantitative PCR (qPCR) Method for Microbial Biofilms

Quantitative PCR (qPCR) enables sensitive and specific quantification of microbial load and gene expression levels within biofilms. This technique is especially valuable for monitoring transcriptional changes in biofilm-associated genes or tracking microbial abundance over time or under treatment conditions.

3.8.1 Overview and Applications

- DNA-based qPCR: Quantify microbial populations or genes (e.g., virulence, resistance, EPS production).
  - Standard curve-based quantification of relative and absolute expression of genes compared to normalization genes.
  - Conventional quantitative real-time PCR (qPCR) amplifies DNA from viable and dead cells, while Viability qPCR is used to eliminate DNA from membrane-compromised cells through treatment with propidium monoazide (PMA) [16].
- RNA-based qPCR (RT-qPCR): Measure gene expression changes under biofilm vs. planktonic states, or in response to stress, antimicrobials, or host interactions.

3.9 Reverse Transcription (for RT-qPCR Only)

If analyzing gene expression, convert RNA to cDNA using high-quality, DNase-treated RNA.

Reverse Transcription Setup (example for 20 μL reaction):

- Template RNA: 500 ng–1 μg (verified for purity and integrity).
- Oligo(dT) or random hexamer primers.
- Reverse transcriptase enzyme (e.g., Super Script IV or iScript).
- dNTPs and buffer (provided with the kit).
- Incubation: Follow manufacturer's guidelines (typically 25 °C–65 °C.)

Aliquot and store cDNA at −20 °C until qPCR.

3.10 qPCR Reaction Setup

Set up reactions in a 96-well plate or strip tubes using a master mix for consistency.

Example 20 μL SYBR Green Reaction Mix

Component	Volume
2× SYBR Green Master Mix	10 μL
Forward Primer (10 μM)	0.4 μL
Reverse Primer (10 μM)	0.4 μL
Template DNA or cDNA	1–2 μL
Nuclease-free water	to 20 μL

Notes:

- Always run samples in *technical triplicates.*
- Include *no-template control (NTC)* and *no-reverse transcription control (NRT)* if doing RT-qPCR.

3.10.1 Thermal Cycling Conditions

Typical SYBR Green cycling protocol:

Step	Temperature	Time	Notes
Initial denaturation	95 °C	2–3 min	Enzyme activation
Denaturation	95 °C	10–15 s	40 cycles
Annealing/ extension	60 °C	30 s	Combined step
Melt curve (optional)	65–95 °C	0.5 °C increments	For product specificity check

Adjust annealing temperature based on primer Tm if needed.

3.10.2 Data Analysis

Ct Value Assessment

- *Ct (cycle threshold):* the cycle at which fluorescence rises above background.
- Lower Ct = higher initial template concentration.

Normalization and Quantification

- Use a *housekeeping gene* (e.g., *16S rRNA*, *rpoD*, *gyrA*) for normalization.
- Choose reference genes with stable expression in both biofilm and planktonic states.

Relative Quantification (ΔΔCt Method)

For gene expression analysis:

- Calculate ΔCt = Ct (target gene) − Ct (reference gene).
- ΔΔCt = ΔCt (treated or biofilm sample) − ΔCt (control or planktonic).
- Fold change = $2^{-\Delta\Delta Ct}$.

Absolute Quantification (Optional)

- Create a standard curve using known concentrations of target gene template.
- Plot Ct values vs. log concentration.
- Use to quantify gene copies per μL.

3.11 Worked-out qPCR Analysis

- *Bacterium: Pseudomonas aeruginosa* (widely used model for biofilm studies).
- *Biofilm-associated target gene: algD*—encodes GDP-mannose dehydrogenase, a key enzyme for alginate biosynthesis and strongly associated with biofilm matrix production in *P. aeruginosa*.

- Sequence Name: algD.
- Amplicon Length: 95.

Tables represent the parameters of forward and reverse primers for *algD* gene designed using IDT primer quest tool. (https://www.idtdna.com/PrimerQuest/Home/Index).

	Start	Stop	Length	Tm	GC%
Forward **GCGATCAAGGACTACGACTTC (Sense)**	132	153	21	62	52.4
Reverse **TCCAGCTCGCGGTAGAT (AntiSense)**	210	227	17	62	58.8

Actual location of the primers in the algD sequence:

Base	Sequence
1	TGGTGCGCAGCACCGTACTGCCGGGCCCGTCAACAACGT GGTGATCCCGCTGATCGAGGACTGCTCGGGCAAGAAG GCCGGGGTCGACTTCGGCGTCGG
101	CACCAACCCCG AATTCCTCCG CGAGAGCACC **GCGATCAAGGA CTAGACTTC** CCGCCGATGACCG TGATCGGCGAAC TGGACAAGCAGA CCGGCGACCTT
201	CTCGAGGAA **ATCTACCGC GAGCTGGA** CGCGCCGATCAT CCGCAAGACCG TCGAGGTC

Reference (housekeeping) gene: rpoD (or gyrA, or 16S rRNA)—choose one validated to be stable between planktonic and biofilm states (we'll use rpoD in the example). Sequence Name: rpoD. Amplicon Length: 109. Primer parameters for rpoD *gene:*

	Start	Stop	Length	Tm	GC %
Forward **TCCTCAGTGGCTATATCGATCC (Sense)**	52	74	22	63	50
Reverse **TCCTCGTCGTCCTTCTCTTT (AntiSense)**	141	161	20	62	50

Actual location of the primers in the rpoD *sequence:*

Base	Sequence
1	TCCTGGCCGAC **TCCTCAGTGGCT** CGATGATGGCAGCC TACAACCGTATCG **ATATCGATCC** TGCCCGCCGAAGA TCGCCGAGGGCGG TCGCCTC TCCGACG
101	GGTGGAGCCGG **AAAGAGAAGGA** AGAAGAAAGCGACGA TCAACC **CGACGAGGA** CAGCAGTGACAGCGA TGAAGGACGA CGACGAAGGC TTCCGCCGACTCG
201	GACGGCGGTC

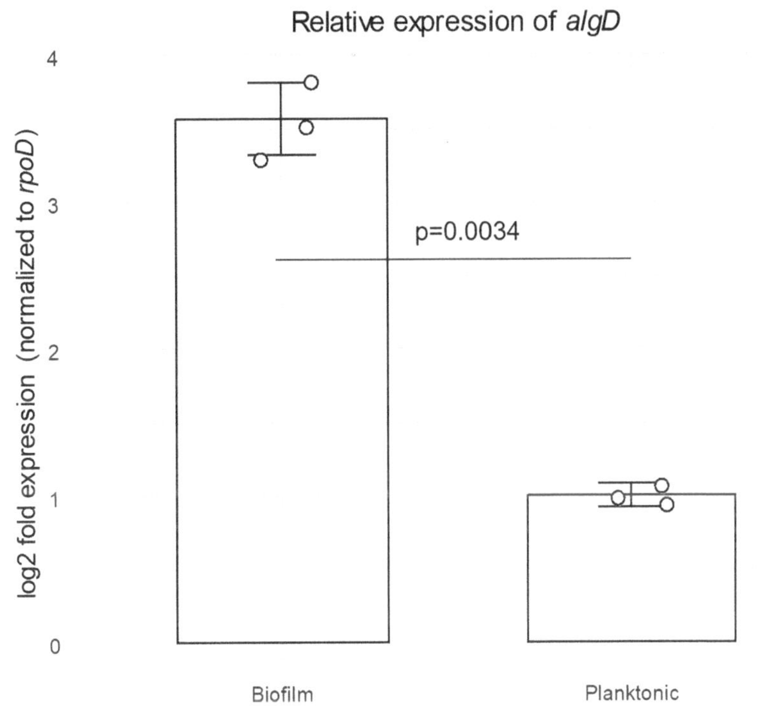

Fig. 2 Relative expression of *algD* in *Pseudomonas aeruginosa*. Bar graph showing the relative expression ($\log_2$ fold change) of *algD*, normalized to *rpoD*, comparing biofilm and planktonic states. Each bar represents the mean of three biological replicates (shown as individual dots), with error bars indicating standard deviation. Statistical significance was determined using Welch's t-test. Graph generated in R

Below is the data obtained from qPCR reactions using *P. aeruginosa* with *algD* (target) and *rpoD* (reference), with three technical replicates per biological sample and the analysis steps shown (Fig. 2).

Example dataset—*Pseudomonas aeruginosa*, target *algD*, ref. *rpoD*

Sample	Condition	Ct (*algD*)—rep1/rep2/rep3	Ct (*rpoD*)—rep1/rep2/rep3
P1	Planktonic	25.1/25.0/25.2	20.2/20.3/20.1
P2	Planktonic	24.8/24.9/24.7	20.1/20.0/20.2
P3	Planktonic	25.3/25.4/25.2	20.4/20.5/20.3
B1	Biofilm	22.9/23.0/22.8	19.9/20.0/19.8
B2	Biofilm	23.1/23.0/23.2	20.0/19.9/20.1
B3	Biofilm	22.7/22.8/22.6	19.8/19.9/19.7

3.11.1 Quick Analysis (Same Pipeline; Numbers Computed)

1. *QC technical replicates*—all tight (SD ≈ 0.1 Ct), keep all wells.
2. *Average technical replicates → one Ct per biological sample.*

Sample	*algD* mean ± SD	*rpoD* mean ± SD	ΔCt = *algD* − *rpoD*
P1	25.10 ± 0.10	20.20 ± 0.10	4.90
P2	24.80 ± 0.10	20.10 ± 0.10	4.70
P3	25.30 ± 0.10	20.40 ± 0.10	4.90
		Planktonic mean ΔCt	**4.833**
B1	22.90 ± 0.10	19.90 ± 0.10	3.00
B2	23.10 ± 0.10	20.00 ± 0.10	3.10
B3	22.70 ± 0.10	19.80 ± 0.10	2.90
		Biofilm mean ΔCt	**3.000**

3. *Group summary (biological replicates).*
   - Planktonic mean ΔCt = 4.833, SD = 0.115.
   - Biofilm mean ΔCt = 3.000, SD = 0.100.
4. *ΔΔCt and fold change.*
   - ΔΔCt = ΔCt_biofilm (B1) − ΔCt_planktonic (average) = 3.000–4.833 = −1.833.
   - Fold change = $2^{(-\Delta\Delta Ct)} = 2^{1.833} \approx 3.6$ → *algD* is ~3.6-fold upregulated in biofilm vs. planktonic.
   - Repeat this for each sample (P1, P2, P3 and B1, B2, B3) and plot a graph as follows.

5. *Statistics (recommended practice).*
   - Test on the ΔCt values across biological replicates (e.g., Welch's t-test or Mann-Whitney if non-normal). With these numbers ($n = 3$ each), the difference is large; report p-value.
   - For reporting fold-change variability, calculate SE on ΔΔCt or report log2(fold) = −ΔΔCt ± SE (as shown previously).

3.11.2 Quality Control and Validation

- Include a melt curve (SYBR assays) to confirm single amplicon specificity.
- Validate primer efficiency (90–110%) using standard curves (serial dilutions).
- Confirm absence of primer-dimers by running a subset of qPCR products on a gel (if needed).

3.12 Selection of qPCR Chemistry: SYBR Green Versus TaqMan

Choosing between SYBR Green and TaqMan probe-based qPCR depends on the study's objectives, budget, target complexity, and desired specificity. Both approaches are widely used in microbial biofilm research, but each has distinct advantages and limitations.

3.12.1 SYBR Green Assays

SYBR Green is a fluorescent dye that binds to any double-stranded DNA (dsDNA), emitting fluorescence proportional to the amount of PCR product generated [17].

Advantages:

- *Cost-effective*—No need for sequence-specific probes.
- *Easy to implement*—Compatible with general-purpose qPCR platforms.
- *Good for screening*—Suitable for high-throughput expression or gene detection studies.

Limitations:

- *Lower specificity*—Binds to all dsDNA, including nonspecific products and primer-dimers.
- *Requires melt curve analysis*—To confirm single, specific amplification.

Recommended Use:

- When working with well-characterized, single-target reactions.
- For comparative gene expression (e.g., ΔΔCt) with validated primers.
- When budget constraints are a concern.

3.12.2 TaqMan Probe-Based Assays

TaqMan assays use a fluorogenic probe (labeled with a reporter and quencher) that binds to a specific sequence within the target amplicon [18]. Fluorescence is released upon probe degradation during amplification.

Advantages:

- *High specificity*—Sequence-specific detection minimizes background.
- *Multiplexing*—Multiple targets can be detected in a single reaction using distinct fluorophores.
- *No melt curve needed*—Easier interpretation of results in complex samples.

Limitations:

- *More expensive*—Custom probe synthesis required.
- *Less flexible*—New targets require new probes.

Recommended Use:

- When working with mixed microbial communities or environmental biofilms.
- For diagnostic applications requiring high specificity.
- In multiplex setups where multiple targets are assessed simultaneously (e.g., species-specific quantification in biofilms).

3.12.3 Decision Summary Table

Criteria	SYBR Green	TaqMan
Specificity	Moderate (requires melt curve)	High (probe-based)
Cost	Low	High
Ease of use	Simple setup	Requires probe design
Multiplexing	Not supported	Supported
Ideal for	Gene expression, initial screening	Diagnostic or confirmatory studies

3.13 Best Practice

For exploratory biofilm gene expression studies, SYBR Green is often sufficient if melt curve analysis confirms specificity. For complex or mixed-species biofilms, or when precise quantification is required (e.g., clinical diagnostics), TaqMan assays offer greater reliability.

3.13.1 Droplet Digital PCR (ddPCR) Method for Microbial Biofilms

Droplet Digital PCR (ddPCR) is an advanced technique for absolute quantification of DNA or cDNA, without the need for standard curves. In biofilm research, ddPCR is particularly advantageous due to its high sensitivity, tolerance to inhibitors, and ability to quantify low-abundance targets within complex microbial communities (*see* **Note 4**) [19].

3.13.2 Principle of ddPCR

In ddPCR, a conventional PCR reaction mix is partitioned into ~20,000 nanoliter-sized droplets, each functioning as an independent micro-reaction. After end-point PCR, droplets are analyzed for fluorescence:

- *Positive droplets* contain amplified target DNA (fluorescent signal).
- *Negative droplets* do not (no signal).

The Poisson distribution is then applied to calculate the absolute copy number of the target per sample volume, independent of Ct values or standard curves.

3.13.3 Advantages of ddPCR in Biofilm Studies

- *Absolute quantification* without external standards.
- *High precision and reproducibility.*
- *Greater tolerance to PCR inhibitors* common in biofilm-derived nucleic acids.
- Ideal for:
  - Low-abundance gene detection.
  - Tracking specific species in multispecies biofilms.
  - Monitoring antibiotic resistance or persistence genes.

ddPCR Reaction Setup (Example for 20 μL Reaction)

Component	Volume
ddPCR Supermix (EvaGreen or probe)	10 μL
Forward primer (10 μM)	0.9 μL
Reverse primer (10 μM)	0.9 μL
Template DNA or cDNA	1–2 μL
Nuclease-free water	to 20 μL

3.13.4 ddPCR Workflow

1. *Prepare Reaction Mix.*
   Combine reagents and mix gently.
2. *Droplet Generation*
   Load the reaction into a droplet generator cartridge along with droplet oil. Generate droplets (manual or automated).

3. *PCR Amplification*
   Transfer droplets to a 96-well plate. Seal and run PCR using recommended ddPCR thermal profile.
4. *Droplet Reading*
   Post-amplification, analyze droplets in a droplet reader. Each droplet is classified as positive or negative based on fluorescence intensity.
5. *Data Analysis*
   Software automatically calculates absolute copy number (copies/µL of reaction) based on Poisson statistics.

Typical ddPCR Thermal Profile

Step	Temperature	Time	Cycles
Enzyme activation	95 °C	10 min	1
Denaturation	94 °C	30 sec	40
Annealing/extension	55–60 °C	60 sec	
Enzyme deactivation	98 °C	10 min	1
Hold	4 °C	∞	

Optimize annealing temperature for primer/probe pair as needed.

3.14 Data Interpretation

- *Positive droplets* exceed fluorescence threshold.
- *Absolute copy number* = copies/µL in reaction (can be normalized to biomass, input volume, etc.).
- *Multiplexing* possible with different probe fluorophores (e.g., FAM, HEX).

3.15 Expected Results

This section outlines the anticipated outcomes from PCR, qPCR, and ddPCR assays when applied to microbial biofilm samples. Given the structural complexity of biofilms, results can vary depending on sample quality, primer design, and amplification method. Below are examples of what successful results typically look like, along with notes on interpretation.

3.16 PCR (Conventional Endpoint PCR)

3.16.1 Gel Electrophoresis Output

- A *single, sharp band* of the expected size (e.g., 100–600 bp) indicates successful amplification.
- A *faint band* may suggest low template concentration or partial inhibition.
- *Multiple bands* or smearing indicate nonspecific amplification or degraded DNA.
- *No band in NTC* (No Template Control) confirms no contamination.

3.16.2 Representative Result Example

Sample	Expected amplicon size	Band on gel?	Interpretation
Biofilm DNA	300 bp	✓	Target gene present
Reference DNA	600 bp	✓	Reference gene present
NTC	–	No band	No contamination

3.17 qPCR (SYBR/TaqMan-Based Real-Time PCR)

3.17.1 Amplification Curves

- A *sigmoidal curve* with a well-defined exponential phase indicates successful amplification.
- *Ct values* (Cycle threshold) typically range from 15 to 35 depending on gene expression or template abundance.
- *Melt curve analysis* (for SYBR): A *single, narrow peak* confirms specific amplification. Multiple peaks suggest primer-dimer or nonspecific products.

3.17.2 Typical Ct Value Interpretation

Sample	Target	Ct value	Interpretation
Biofilm (target gene)	*lasR*	21.5	High expression
Planktonic (same gene)	*lasR*	28.9	Lower expression
NTC	–	Undetermined	No amplification, clean setup
No RT control	–	Undetermined	RNA-specific signal confirmed

3.17.3 ΔΔCt Fold Change Example

If normalized to housekeeping gene (*16S*) and planktonic as baseline:

$\Delta\text{Ct (biofilm)} = 21.5–16.0 = 5.5.$

$\Delta\text{Ct (planktonic)} = 28.9–16.5 = 12.4.$

$\Delta\Delta\text{Ct} = 5.5–12.4 = -6.9.$

Fold change $= 2^{6.9} \approx$ *122-fold increase in expression* in biofilm

3.18 ddPCR (Droplet Digital PCR)

3.18.1 Droplet Fluorescence Output

- Positive droplets are clearly separated from negative droplets on a *1D fluorescence amplitude plot.*
- *Absolute quantification* is provided in *copies/μL of reaction.*
- A well-partitioned assay yields *>10,000 total droplets*, with ≥95% confidence interval.

3.18.2 Representative Output Table

Sample	Positive droplets	Total droplets	Copies/ μL	Interpretation
Biofilm DNA (*mucA*)	5326	19,870	1130	High abundance
Planktonic DNA (*mucA*)	972	20,150	200	Low abundance
NTC	0	19,960	0	No contamination
Sample with EPS contamination	600	12,340	128	Possibly inhibited; consider dilution or cleanup

3.18.3 Rain (Ambiguous Droplets)

- Observed between positive and negative clusters. Typically indicates:
  - Suboptimal annealing temperature.
  - Degraded probe or off-target amplification.
  - Adjust reaction conditions or redesign primers.

4 Conclusion

This chapter outlines up-to-date technologies for practical and rigorously validated PCR-based methodologies for the analysis of microbial biofilms, encompassing conventional PCR and nested PCR for detection, qPCR for quantitative and functional assessment, and ddPCR for absolute and high-resolution measurement of biofilm-associated targets. By addressing biofilm-specific challenges such as low biomass, extracellular polymeric matrices, and microbial heterogeneity, these approaches enable accurate detection, quantification, and molecular characterization of biofilm-forming microorganisms and their associated genes. The workflows, troubleshooting strategies, and comparative guidance provided here are intended to support reproducible experimental design and informed method selection across diverse biofilm systems. Collectively, these molecular tools will continue to play a critical role in advancing our understanding of biofilm biology and in supporting the development of effective strategies to monitor, prevent, and control biofilm-associated infections and processes.

5 Notes

1. Due to the complex nature of microbial biofilms, including the presence of extracellular polymeric substances (EPS), inhibitors, and mixed microbial populations, PCR reactions can be prone to technical issues. Below are the common problems and their potential solutions:

Problem	Possible cause	Recommended solution
No amplification (no band)	• PCR inhibitors in DNA extract (e.g., EPS, humic acids) • DNA degraded or too low in concentration • Incorrect thermocycling conditions • Faulty primers or expired reagents	• Clean up DNA using column-based purification or ethanol precipitation • Verify DNA integrity and concentration via gel or Qubit • Increase template volume or use BSA (0.1 mg/mL) in the reaction • Optimize annealing temperature using gradient PCR • Check primer design and storage
Smearing on gel	• Too much template • Nonspecific amplification • Poor primer specificity or degraded primers	• Reduce DNA input • Use hot-start polymerase • Redesign primers or increase annealing temperature • Decrease cycle number
Multiple nonspecific bands	• Low primer specificity • Suboptimal annealing temperature • Contamination	• Redesign primers • Increase annealing temperature by 2–5 °C • Use touchdown PCR protocol • Use fresh reagents and practice sterile technique
Primer-dimer formation	• Complementarity at 3′ ends • Excessive primer concentration	• Redesign primers to avoid 3′ complementarity • Reduce primer concentration to 0.1–0.25 μM
Weak or faint bands	• Low template concentration • Suboptimal Mg^{2+} concentration • Inefficient primer binding	• Increase template volume slightly • Adjust $MgCl_2$ concentration in 0.5 mM increments • Confirm primer integrity and adjust Tm
Amplification in NTC (negative control)	• Contamination of reagents or workspace • Primer-dimer formation	• Replace reagents (especially water and primers) • Prepare master mix in a clean area using filter tips • Use separate pipettes for template and master mix setup

Tips for PCR with Biofilm-Derived DNA:

- Always include a no-template control (NTC) to check for contamination.
- Include an internal control gene (e.g., 16S rRNA) to verify PCR efficiency.
- When working with environmental or clinical biofilms, diluting the template DNA 1:10 may reduce the concentration of inhibitors while still retaining amplifiable DNA.

2.

Issue	Cause	Solution
No amplification	Inhibitors or degraded template	Re-purify template, dilute sample 1:10
High Ct values	Low input DNA/cDNA or poor primer efficiency	Verify template quality, redesign primers
Inconsistent replicates	Pipetting error or uneven mixing	Use master mix, calibrate pipettes
Multiple peaks in melt curve	Nonspecific amplification	Increase annealing temp, optimize primers
Amplification in NTC	Contamination	Replace reagents, use filtered tips and clean workspace

3.

Issue	Cause	Solution
Few or no positive droplets	Low template concentration, inhibitors	Concentrate DNA, purify further
High number of positive droplets in NTC	Contamination	Replace reagents, clean workspace
Low total droplet count	Oil contamination or poor pipetting	Ensure clean oil and proper droplet generation
Rain (ambiguous droplets)	Suboptimal annealing temp, probe degradation	Optimize temp, check probe integrity

4. When to Choose ddPCR

Use case	Recommendation
Very low copy targets in biofilms	ddPCR
Inhibitor-rich samples (e.g., EPS-heavy matrix)	ddPCR
Absolute quantification without standard curve	ddPCR
Gene expression or high-throughput screening	Prefer qPCR (SYBR/ TaqMan)
Budget-sensitive studies	Prefer qPCR

Competing Interest

The author reports no financial conflicts related to the subject matter of this chapter.

References

1. Shree P, Singh CK, Sodhi KK, Surya JN, Singh DK (2023) Biofilms: understanding the structure and contribution towards bacterial resistance in antibiotics. Med Microecol 16: 100084. https://doi.org/10.1016/j.medmic.2023.100084
2. Zhao A, Sun J, Liu Y (2023) Understanding bacterial biofilms: from definition to treatment strategies. Front Cell Infect Microbiol 13: 1137947. https://doi.org/10.3389/fcimb.2023.1137947
3. Mirghani R, Saba T, Khaliq H, Mitchell J, Do L, Chambi L, Diaz K, Kennedy T, Alkassab K, Huynh T, Elmi M, Martinez J, Sawan S, Rijal G (2022) Biofilms: formation, drug resistance and alternatives to conventional approaches. AIMS Microbiol 8(3):239–277. https://doi.org/10.3934/microbiol.2022019
4. Nett JE, Andes DR (2015) Fungal biofilms: In vivo models for discovery of anti-biofilm drugs. Microbiol Spectrum 3(4):MB-0008-2014. https://doi.org/10.1128/microbiolspec.MB-0008-2014
5. Kumaresan V, Pasupuleti M, Paray BA, Al-Sadoon MK, Arockiaraj J (2019) Gene profiling of antimicrobial peptides, complement factors and MHC molecules from the skin transcriptome of *Channa striatus* and its expression pattern during *Aeromonas hydrophila* infection. Fish Shellfish Immunol 84:48–55. https://doi.org/10.1016/j.fsi.2018.09.061
6. De Silva L, Heo GJ (2022) Biofilm formation of pathogenic bacteria isolated from aquatic animals. Arch Microbiol 205(1):36. https://doi.org/10.1007/s00203-022-03332-8
7. Sannasimuthu A, Kumaresan V, Anilkumar S, Pasupuleti M, Ganesh M-R, Mala K, Paray BA, Al-Sadoon MK, Albeshr MF, Arockiaraj J (2019) Design and characterization of a novel *Arthrospira platensis* glutathione oxidoreductase-derived antioxidant peptide GM15 and its potent anti-cancer activity via caspase-9 mediated apoptosis in oral cancer cells. Free Radic Biol Med 135:198–209. https://doi.org/10.1016/j.freeradbiomed.2019.03.006
8. Giaouris E (2025) Comparing gene expression between planktonic and biofilm cells of food-borne bacterial pathogens through RT-qPCR. Methods Mol Biol 2852:143–158. https://doi.org/10.1007/978-1-0716-4100-2_10
9. Magalhães AP, França Â, Pereira MO, Cerca N (2019) RNA-based qPCR as a tool to quantify and to characterize dual-species biofilms. Sci Rep 9(1):13639. https://doi.org/10.1038/s41598-019-50094-3
10. Nahum Y, Muhvich J, Morones-Ramirez JR, Casillas-Vega NG, Zaman MH (2025) Biofilms as potential reservoirs of antimicrobial

resistance in vulnerable settings. Front Public Health 13:1568463. https://doi.org/10.3389/fpubh.2025.1568463

11. Tuytschaever T, Raes K, Sampers I (2025) Biofilm detection in the food industry: challenges in identifying biofilm EPS markers and analytical techniques with insights for *Listeria monocytogenes*. Int J Food Microbiol 432:111091. https://doi.org/10.1016/j.ijfoodmicro.2025.111091
12. Roux D, Cywes-Bentley C, Zhang Y-F, Pons S, Konkol M, Kearns DB, Little DJ, Howell PL, Skurnik D, Pier GB (2015) Identification of poly-N-acetylglucosamine as a major polysaccharide component of the *Bacillus subtilis* biofilm matrix. J Biol Chem 290(31): 19261–19272. https://doi.org/10.1074/jbc.M115.648709
13. Crawford EC, Singh A, Metcalf D, Gibson TW, Weese SJ (2014) Identification of appropriate reference genes for qPCR studies in *Staphylococcus pseudintermedius* and preliminary assessment of icaA gene expression in biofilm-embedded bacteria. BMC Res Notes 7:451. https://doi.org/10.1186/1756-0500-7-451
14. Zegaer BH, Ioannidis A, Babis GC, Ioannidou V, Kossyvakis A, Bersimis S, Papaparaskevas J, Petinaki E, Pliatsika P, Chatzipanagiotou S (2014) Detection of bacteria bearing resistant biofilm forms by using universal and specific PCR is still unhelpful in the diagnosis of periprosthetic joint infections. Front Med 1:30. https://doi.org/10.3389/fmed.2014.00030
15. Niedzielski A, Chmielik LP, Stankiewicz T (2021) The formation of biofilm and bacteriology in otitis media with effusion in children: a prospective cross-sectional study. Int J Environ Res Public Health 18(7):3555. https://doi.org/10.3390/ijerph18073555
16. Schwendener S, Flury M, Jenzer J, Thurnheer T, Karygianni L (2025) PMA-qPCR to quantify viable cells in multispecies oral biofilm after disinfectant treatments. Biofilms 9:100281. https://doi.org/10.1016/j.bioflm.2025.100281
17. França A, Freitas AI, Henriques AF, Cerca N (2012) Optimizing a qPCR gene expression quantification assay for *Staphylococcus epidermidis* biofilms: a comparison between commercial kits and a customized protocol. PLoS One 7(5):e37480. https://doi.org/10.1371/journal.pone.0037480
18. Gionchetta G, Snead D, Semerad S, Beck K, Pruden A, Bürgmann H (2023) Dynamics of antibiotic resistance markers and *Escherichia coli* invasion in riverine heterotrophic biofilms facing increasing heat and flow stagnation. Sci Total Environ 893:164658. https://doi.org/10.1016/j.scitotenv.2023.164658
19. Grudlewska-Buda K, Skowron K, Gospodarek-Komkowska E (2020) Comparison of the intensity of biofilm formation by *Listeria monocytogenes* using classical culture-based method and digital droplet PCR. AMB Express 10(1): 75. https://doi.org/10.1186/s13568-020-01007-5

Chapter 11

Characterization of Biofilm-Associated Microbial Communities Using 16S rRNA Gene Sequencing

Abirami K

Abstract

Biofilms represent a dominant and ecologically significant mode of microbial growth, characterized by surface-associated communities embedded within an extracellular polymeric substance matrix. Accurate characterization of biofilm-associated microbial communities is essential for understanding their structure, diversity, and functional potential. However, molecular analysis of biofilms poses unique technical challenges due to low biomass, matrix-associated inhibitors, and the need for efficient biofilm disruption.

This chapter describes a standardized and reproducible methodology for profiling biofilm-associated bacterial communities using 16S ribosomal RNA (rRNA) gene amplicon sequencing. The workflow encompasses biofilm cultivation under controlled conditions, careful harvesting of surface-attached biomass, and biofilm-optimized DNA extraction incorporating mechanical and chemical lysis strategies. Amplification of targeted hypervariable regions of the 16S rRNA gene is performed using high-fidelity polymerase chain reaction, followed by library preparation and high-throughput sequencing.

Subsequent bioinformatics analysis includes quality filtering, sequence denoising, taxonomic assignment, and diversity assessment using curated reference databases. The described protocol is adaptable to a variety of biofilm models and experimental designs, providing a robust framework for comparative analysis of microbial community composition across different biofilm conditions. This methodological approach enables high-resolution insight into biofilm microbiota and supports ecological, clinical, and industrial biofilm research.

Key words DNA extraction from biofilms, Microbial community profiling, Amplicon sequencing, Microbial diversity, High-throughput sequencing, Biofilm-associated microbial communities

1 Introduction

Microbial biofilms are ubiquitous and represent a dominant form of microbial life in natural, engineered, and clinical environments. In biofilms, microorganisms adhere to surfaces and are encased within a self-generated extracellular polymeric substance (EPS) matrix made up of polysaccharides, proteins, lipids, and extracellular DNA. The EPS matrix not only provides structural integrity but also creates a protective microenvironment that enhances microbial

Abirami K (ed.), *Microbial Biofilm Research: Comprehensive Protocols and Best Practices*, Springer Protocols Handbooks, https://doi.org/10.1007/978-1-0716-5535-1_11,

tolerance to environmental stressors, antimicrobial agents, and host immune responses. This spatially structured lifestyle promotes complex interactions, including metabolic cooperation and signaling among community members. Because of these characteristics, biofilms are critically involved in chronic infections, biogeochemical processes, industrial fouling, wastewater treatment, and contamination of food and biomedical devices [1].

Understanding the microbial composition and dynamics of biofilms is essential for deciphering their ecological functions, pathogenic potential, and responses to experimental or environmental changes. Traditional culture-dependent methods capture only a fraction of microbial diversity, as many biofilm-associated taxa are slow-growing or non-culturable under standard laboratory conditions. Consequently, culture-independent molecular approaches have become foundational in biofilm microbiology, allowing direct analysis of microbial communities without the need for cultivation. Sequencing of the 16S ribosomal RNA (rRNA) gene is among the most widely used of these methods, providing a robust framework for taxonomic profiling of bacterial and archaeal populations [2].

The 16S rRNA gene contains conserved regions that enable broad-range primer binding, interspersed with hypervariable regions that confer taxonomic discrimination across diverse microbial taxa. High-throughput sequencing of amplified 16S rRNA gene fragments enables detailed characterization of community structure and diversity, making it a cornerstone technique in microbial ecology and biofilm research. Recent developments in sequencing technologies, including long-read platforms (e.g., Oxford Nanopore and PacBio), have further enhanced taxonomic resolution and extended the capacity to profile complex biofilm communities with improved species-level discrimination.

Despite its strengths, applying 16S rRNA gene sequencing to biofilm samples presents methodological challenges. Biofilms often yield low and spatially heterogeneous biomass, particularly in early developmental stages or in nutrient-limited environments. The EPS matrix can hinder efficient cell lysis and co-purify inhibitory substances such as polysaccharides and humic acids, which interfere with DNA extraction and downstream PCR amplification. Mechanical detachment and disruption methods that are necessary to recover biofilm biomass can also introduce bias or cause DNA shearing if not carefully controlled. These technical hurdles necessitate specific protocol considerations at each stage of the workflow—from sample collection and biofilm detachment to DNA extraction, inhibitor removal, and PCR amplification [3].

To address these issues, optimized strategies combining mechanical lysis (e.g., bead beating) with efficient inhibitor removal chemistries are critical for obtaining high-quality, amplifiable DNA from biofilm matrices. Additionally, careful selection of primer sets, amplification conditions, and sequencing platforms is vital to minimize bias and ensure reproducible community profiling. The protocol presented here describes a standardized, adaptable workflow for 16S rRNA gene sequencing of biofilm-associated microbial communities, suitable for a range of model systems, including static biofilms on microtiter plates or glass surfaces and dynamic systems such as flow cells or reactor coupons. Emphasis is placed on contamination control, quality assurance, and methodological consistency to generate reliable and interpretable sequencing data [Characterization of Biofilm-Associated Microbial Communities Using 16S rRNA Gene Sequencing] [4].

By following this workflow, researchers can gain high-resolution insights into biofilm microbial communities, enabling comparative analyses across experimental conditions, temporal stages, or treatment regimes. Such analyses are critical for advancing our understanding of biofilm ecology, function, and their broader implications in clinical, environmental, and industrial microbiology.

2 Materials

2.1 Biofilm Cultivation Materials

- Sterile culture media (e.g., LB broth, TSB, M9, or appropriate medium).
- Sterile microtiter plates (24-well/96-well) or culture tubes.
- Glass coverslips or biofilm coupons (stainless steel, plastic, or glass).
- Incubator (with temperature control).
- Shaker/incubator (optional, for dynamic conditions).
- Sterile pipette tips and micropipettes.
- Laminar airflow cabinet.
- Sterile phosphate-buffered saline (PBS).

2.2 Biofilm Harvesting Materials

- Sterile cell scrapers or pipette tips.
- Sterile microcentrifuge tubes (1.5 mL/2.0 mL).
- Sonicator (bath or probe, low power).
- Vortex mixer.
- Refrigerated centrifuge.
- Ice bucket.

2.3 DNA Extraction Materials (Biofilm-Optimized)

- Biofilm-specific DNA extraction kit

  (e.g., Qiagen DNeasy PowerBiofilm Kit / PowerSoil Pro / ZymoBIOMICS DNA Miniprep Kit).
- Bead-beating tubes with glass or ceramic beads.
- Bead beater or homogenizer.

- Proteinase K (if required).
- RNase A (optional).
- Nuclease-free water.
- Ethanol (molecular biology grade).

2.4 DNA Quantification and Quality Assessment

- NanoDrop spectrophotometer.
- Qubit fluorometer with dsDNA assay kit.
- Agarose.
- Gel electrophoresis unit.
- DNA ladder.
- Gel documentation system.

2.5 16S rRNA Gene Amplification Materials

- High-fidelity PCR master mix.
- 16S rRNA gene primers (e.g., V3–V4 region primers).
- PCR tubes or plates.
- Thermal cycler.
- PCR-grade water.
- Positive and negative control DNA.

2.6 Library Preparation Materials

- AMPure XP magnetic beads.
- Magnetic stand.
- Indexing primers (Illumina-compatible).
- PCR purification kit (optional).
- Microcentrifuge tubes.
- Pipettes and sterile tips.

2.7 Sequencing Materials

- Illumina sequencing platform (e.g., MiSeq).
- Sequencing reagents and flow cell.
- Library denaturation and dilution reagents.
- PhiX control library (optional).

2.8 Bioinformatics and Data Analysis

- Computer workstation.
- Bioinformatics software (QIIME2/DADA2/Mothur).
- Reference databases (SILVA/Greengenes).
- Statistical analysis software (R/Python).

2.9 General Laboratory Consumables

- Gloves, lab coats, face masks.
- Disinfectants (70% ethanol).
- Autoclaved waste disposal bags.
- Marker pens and labeling tape.

2.10 Optional (Recommended for Quality Control)

- Mock microbial community standard.
- DNA extraction negative control.
- PCR no-template control.

3 Methods

3.1 Biofilm Cultivation

Steps

1. Select an appropriate biofilm model (static or dynamic).
2. Prepare sterile growth medium suitable for the target microorganisms.
3. Inoculate the medium with the microbial culture or environmental sample.
4. Dispense the inoculated medium into sterile microtiter plates, culture tubes, or onto glass/coupon surfaces.
5. Incubate under controlled temperature and atmospheric conditions for 24–72 h or until biofilm maturation.
6. Carefully replace the spent medium at defined intervals to remove planktonic cells without disturbing the biofilm (*see* **Note 1**).
7. Maintain aseptic conditions throughout the cultivation period (*see* **Note 2**) [1].

3.2 Biofilm Harvesting

1. Remove planktonic culture gently using a pipette (*see* **Note 3**).
2. Wash the biofilm surface 1–2 times with sterile phosphate-buffered saline (PBS).
3. Add sterile PBS to the biofilm surface.
4. Detach the biofilm using a sterile scraper or pipette tip.
5. Transfer the biofilm suspension into sterile microcentrifuge tubes.
6. Sonicate briefly at low power (30–60 s) to disrupt the EPS matrix (*see* **Notes 4** and **5**).
7. Centrifuge at 10,000× g for 10 min.
8. Discard the supernatant and retain the pellet for DNA extraction (Fig. 1).

3.3 DNA Extraction (Biofilm-Optimized)

1. Resuspend the biofilm pellet in the lysis buffer provided with a biofilm-optimized DNA extraction kit (*See* **Note 6**).
2. Add bead-beating matrix to the tube.
3. Perform mechanical lysis using a bead beater as per kit instructions.
4. Incubate with enzymatic lysis reagents (if provided) to enhance cell wall disruption.

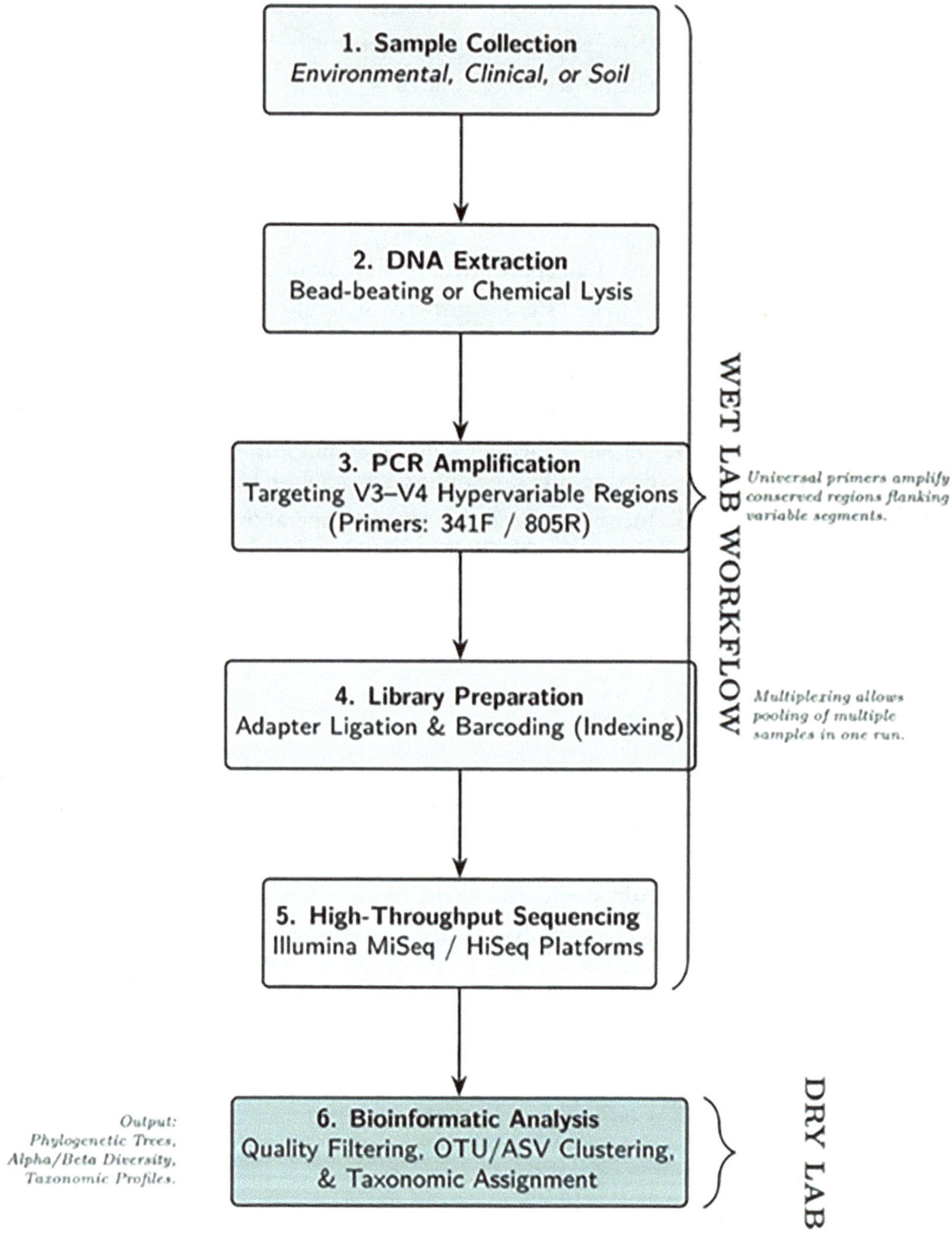

Fig. 1 Workflow for 16S rRNA gene sequencing of biofilm samples.Schematic representation of the stepwise methodology used for microbial community profiling of biofilm-associated samples

5. Remove inhibitors using the supplied inhibitor-removal solution.
6. Bind DNA to a silica column or magnetic beads.
7. Wash the bound DNA to remove residual contaminants.
8. Elute purified DNA in nuclease-free water or elution buffer (*see* **Note** 7).
9. Quantify DNA concentration and assess purity using spectrophotometry or fluorometry.

3.4 16S rRNA Gene Amplification

1. Prepare PCR master mix containing high-fidelity DNA polymerase, primers, dNTPs, and buffer (*see* **Note 8**).
2. Add extracted DNA as template.
3. Include negative (no-template) control (*see* **Note 9**).
4. Perform PCR targeting a selected hypervariable region (e.g., V3–V4).
5. Use optimized cycling conditions (typically 25–30 cycles).
6. Verify PCR amplification by agarose gel electrophoresis.
7. Proceed only with samples showing correct amplicon size [2].

3.5 Library Preparation

Steps

1. Purify PCR amplicons using magnetic beads (*see* **Note 10**).
2. Attach sequencing adapters and index sequences via index PCR or ligation.
3. Perform post-indexing cleanup to remove free adapters.
4. Assess library size distribution and concentration.
5. Normalize individual libraries to equal concentrations.
6. Pool indexed libraries to generate a final sequencing library (*see* **Note 11**) [5].

3.6 Sequencing

Steps

1. Prepare the pooled library according to the sequencing platform protocol.
2. Load the library onto the sequencing instrument (e.g., Illumina MiSeq).
3. Perform paired-end sequencing.
4. Include sequencing controls such as mock communities and negative controls.
5. Generate raw sequence data in FASTQ format (*see* **Note 12**) [6].

3.7 Bioinformatics Analysis

Steps

1. Import raw FASTQ files into a bioinformatics pipeline (e.g., QIIME2 or DADA2) (*see* **Notes 13** and **14**).
2. Perform quality filtering and trimming of low-quality reads.
3. Merge paired-end reads.
4. Remove chimeric sequences.
5. Generate amplicon sequence variants (ASVs) or OTUs.
6. Assign taxonomy using a reference database (e.g., SILVA).
7. Calculate alpha and beta diversity metrics (*see* **Note 15**).
8. Perform statistical and comparative analyses.
9. Visualize microbial community composition using plots and charts [7].

4 Notes

1. Dynamic systems (e.g., flow cells or biofilm reactors) may be used to simulate shear stress and continuous nutrient flow.
2. Aseptic conditions must be strictly maintained during biofilm cultivation, harvesting, and DNA extraction to prevent external contamination that may interfere with sequencing results.
3. Planktonic cells should be removed carefully before biofilm harvesting to ensure that only surface-attached biofilm microorganisms are analyzed.
4. Biofilm disruption should be gentle and controlled. Excessive sonication or mechanical force can cause DNA shearing and affect downstream PCR and sequencing quality.
5. Biofilm samples often yield low DNA concentrations; therefore, unnecessary sample dilution should be avoided, and sensitive DNA quantification methods should be used.
6. Efficient inhibitor removal is critical, as EPS components such as polysaccharides and humic substances can inhibit PCR amplification.
7. High-fidelity DNA polymerase should be used during 16S rRNA gene amplification to minimize sequencing errors and amplification bias.
8. The number of PCR cycles should be optimized (generally not exceeding 30 cycles) to avoid chimera formation and over-amplification.
9. Negative controls must be included during DNA extraction and PCR to detect reagent or environmental contamination.
10. Amplicon size should be confirmed by agarose gel electrophoresis before proceeding to library preparation.

11. Libraries must be accurately quantified and normalized prior to sequencing to ensure balanced sample representation.
12. Sequencing controls, such as mock communities, are recommended to assess sequencing accuracy and pipeline performance.
13. Bioinformatics parameters should be standardized across samples to allow reliable comparison of microbial community data.
14. All samples and reagents should be properly labeled and documented to ensure traceability and reproducibility of results.
15. Data should be backed up and stored securely, and raw sequencing files should be retained for re-analysis if required.

Competing Interest

The author confirms that there are no financial interests that could be perceived as influencing the content of this chapter.

References

1. Coenye T, Ahonen M, Anderson S, Cámara M, Chundi P, Fields M, Foidl I, Gnimpieba EZ, Griffin K, Hinks J, Loka AR, Lushbough C, Mac Phee C, Nater N, Raval R, Slater-Jefferies J, Teo P, Wilks S, Yung M, Biofilm Priority Questions Exercise Participants; Webb JS (2024) Global challenges and microbial biofilms: identification of priority questions in biofilm research, innovation and policy. Biofilms 8:100210. https://doi.org/10.1016/j.bioflm.2024.100210
2. Regueira-Iglesias A, Balsa-Castro C, Blanco-Pintos T, Tomás I (2023) Critical review of 16S rRNA gene sequencing workflow in microbiome studies: from primer selection to advanced data analysis. Mol Oral Microbiol 38(5):347–399. https://doi.org/10.1111/omi.12434
3. Esberg A, Fries N, Haworth S et al (2024) Saliva microbiome profiling by full-gene 16S rRNA Oxford Nanopore technology versus Illumina MiSeq sequencing. NPJ Biofilms Microbiomes 10:149. https://doi.org/10.1038/s41522-024-00634-1
4. Aja-Macaya P, Conde-Pérez K, Trigo-Tasende N et al (2025) Nanopore full length 16S rRNA gene sequencing increases species resolution in bacterial biomarker discovery. Sci Rep 15:26486. https://doi.org/10.1038/s41598-025-10999-8
5. Klindworth A, Pruesse E, Schweer T, Peplies J, Quast C, Horn M, Glöckner FO (2013) Evaluation of general 16S ribosomal RNA gene PCR primers for classical and next-generation sequencing-based diversity studies. Nucleic Acids Res 41(1):e1. https://doi.org/10.1093/nar/gks808
6. Callahan BJ, McMurdie PJ, Rosen MJ, Han AW, Johnson AJA, Holmes SP (2016) *DADA2:* High-resolution sample inference from Illumina amplicon data. Nat Methods 13:581–583. https://doi.org/10.1038/nmeth.3869
7. Flemming H-C, Wingender J (2010) The biofilm matrix. Nat Rev Microbiol 8(9):623–633. https://doi.org/10.1038/nrmicro2415

Chapter 12

Metagenomic Approaches for the Detection and Characterization of Biofilms

Abirami K

Abstract

Biofilms are complex microbial ecosystems in which diverse microorganisms organize into structured communities and become enclosed within a self-produced extracellular polymeric substance (EPS) matrix. These resilient structures attach to a wide range of surfaces and play major roles in persistent clinical infections, industrial fouling, and environmental contamination. Conventional culture-based microbiology often provides an incomplete picture of biofilm composition because many biofilm-dwelling organisms grow slowly, rely on interactions with neighboring species, or exist in dormant, non-culturable states. Metagenomic next-generation sequencing (mNGS) overcomes these limitations by enabling direct, culture-independent analysis of DNA or RNA extracted from biofilm samples. This high-throughput approach captures the full spectrum of microbial life, bacteria, archaea, fungi, and viruses, and detects both dominant and low-abundance community members. Beyond taxonomic profiling, mNGS uncovers functional genes associated with biofilm development and survival, including those involved in EPS formation, quorum sensing, antimicrobial resistance, virulence, metabolic pathways, and horizontal gene transfer. Through these capabilities, mNGS provides a comprehensive and high-resolution understanding of the diversity, functionality, and ecological impact of microbial biofilms.

Key words Microbial diversity, Culture-independent analysis, Taxonomic profiling, Functional genomics, Antibiotic resistance

1 Introduction

Biofilms are structured microbial communities encased in a self-produced extracellular polymeric substance (EPS) matrix, which adheres to biotic or abiotic surfaces. These biofilm communities are highly heterogeneous and often consist of multiple microbial species, interacting dynamically to create a stable, resilient environment. Biofilms are implicated in a variety of clinical, environmental, and industrial challenges, such as chronic infections, medical device contamination, industrial fouling, and water distribution system contamination [1]. Traditional microbiological methods rely primarily on culturing techniques to detect and identify microbes.

Abirami K (ed.), *Microbial Biofilm Research: Comprehensive Protocols and Best Practices*, Springer Protocols Handbooks, https://doi.org/10.1007/978-1-0716-5535-1_12,

However, biofilm-forming microorganisms are notoriously difficult to culture in the laboratory due to their slow growth, dependence on interspecies interactions, presence in dormant states (e.g., viable but non-culturable cells), resistance to conventional laboratory media, and environmental stresses. As a result, culture-based approaches often fail to detect many biofilm-associated organisms or underestimate their diversity and functional potential [2, 3].

1.1 Role of Metagenomic Next-Generation Sequencing (mNGS)

Metagenomic next-generation sequencing (mNGS) provides an advanced, unbiased approach for the detection and characterization of complex microbial communities directly from biofilm samples, without the need for prior cultivation. The technique involves the extraction of total DNA or RNA from the biofilm matrix, followed by high-throughput sequencing, yielding millions of reads that represent the entire microbial population and its functional potential. Metagenomic next-generation sequencing (mNGS) offers several key advantages for studying biofilms [4]. It enables culture-independent detection of both culturable and non-culturable microorganisms, providing a comprehensive view of microbial diversity. The high sensitivity and sequencing depth allow identification of low-abundance species and rare functional genes. mNGS provides detailed species-level taxonomic profiling across bacteria, archaea, fungi, and viruses. It also enables functional genomic analysis, identifying genes related to EPS production, quorum sensing, antibiotic resistance, virulence, and metabolic pathways crucial for biofilm persistence. Additionally, mNGS facilitates the detection of horizontal gene transfer, revealing mobile genetic elements involved in resistance and virulence. Finally, it offers semiquantitative data, allowing assessment of microbial abundance and the dominance of key biofilm species [5].

2 Materials

2.1 Sample Collection and Preservation

- Swabs (nylon-flocked, sterile).
- Scrapers or scalpels.
- Sterile pipettes and tips.
- Sterile sonication bath or sonicator probe (for device biofilms).
- Phosphate-buffered saline (PBS) or another sterile buffer.
- Nucleic acid stabilization solutions (e.g., RNA later, DNA/RNA Shield).
- Sterile cryovials or tubes.
- Ice, dry ice, −80 °C freezers for storage.
- Sample logs, labels, barcode system.

2.2 Nucleic Acid Extraction

- Bead-beating tubes with beads (0.1–0.5 mm for bacteria, 0.5–1 mm for fungi).
- Detergents: SDS, Triton X-100.
- Enzymes: lysozyme, proteinase K, chitinase, lyticase.
- Sonicator or mechanical homogenizer.
- Qiagen DNeasy Blood & Tissue Kit, Power Water, or equivalent.
- RNA extraction kits (e.g., RNeasy Mini Kit).
- RNase/DNase-free tubes and tips.
- Filtered pipette tips.
- Nanodrop spectrophotometer, Qubit fluorometer.
- Agarose gel electrophoresis materials.

2.3 Library Preparation

- 16S rRNA primers (e.g., V3-V4 region).
- ITS primers for fungi.
- PCR reagents (Taq polymerase, dNTPs, buffers).
- PCR purification kits.
- DNA fragmentation kits or sonicator.
- End-repair and adapter ligation reagents.
- Library amplification kits (e.g., Illumina TruSeq, Nextera).
- RNA cleanup kits.
- rRNA depletion kits (bacterial and fungal).
- Reverse transcription kits.
- RNA fragmentation reagents.
- Bioanalyzer or TapeStation.
- qPCR kits for library quantification.

2.4 Sequencing Platforms and Consumables

- Illumina (MiSeq, NextSeq, NovaSeq) for short reads.
- Oxford Nanopore (MinION, GridION) or PacBio Sequel for long reads.
- Flow cells or sequencing chips.
- Sequencing reagents and buffers specific to platform.
- Barcodes/adapters for multiplexing.

2.5 Controls

- *Negative controls:* Extraction blanks.

  No-template controls (NTC) for PCR.
- *Positive controls:* Mock microbial communities (bacterial and fungal).

  Synthetic DNA/RNA spike-ins for calibration.

2.6 Bioinformatics and Computational Materials

- High-performance computing (HPC) cluster or workstation.
- Sufficient storage for raw and processed sequencing data.
- Quality control: FastQC, Fastp, Trim Galore.
- Host read removal: KneadData, DeconSeq.
- Taxonomic profiling: Kraken2, MetaPhlAn, Centrifuge.
- Functional annotation: Prokka, RFW, CARD (AMR), VFDB (virulence), KEGG.
- Workflow management: Snakemake, Nextflow, Galaxy.
- RefSeq, GTDB (bacteria), UNITE (fungi).
- AMR and virulence factor databases.

2.7 General Laboratory Equipment

- Biosafety cabinet (BSL-2).
- Centrifuges (microcentrifuge and refrigerated).
- Vortex mixer.
- Pipettes (single and multichannel).
- Incubators for culture-based controls (optional).
- Autoclave for decontamination.
- RNase/DNase-free work surfaces and consumables.

3 Methods

3.1 Experimental Design Overview

This study was designed to detect, identify, and functionally characterize bacterial and fungal communities involved in biofilm formation using metagenomic next-generation sequencing (mNGS). Depending on the scientific question, different analytical focuses are considered, including (i) taxonomic detection and identification, (ii) relative abundance profiling, (iii) functional gene characterization such as antimicrobial resistance (AMR) and virulence determinants, and (iv) activity profiling through RNA-based approaches. To achieve these aims, sequencing strategies were selected according to the desired resolution: targeted amplicon sequencing (16S rRNA for bacteria, internal transcribed spacer [ITS] for fungi) for initial community profiling, shotgun metagenomics for comprehensive taxonomic and functional assessment, and metatranscriptomics to investigate transcriptionally active biofilm components.

3.2 Sample Collection and Preservation

Biofilm samples should be collected using methods that preserve structural integrity and minimize contamination (e.g., sterile swabbing, scraping, or sonication of devices). Immediate stabilization and maintenance of the cold chain are essential to protect microbial DNA and RNA, including extracellular DNA within the biofilm matrix. Detailed metadata such as collection site, time, storage

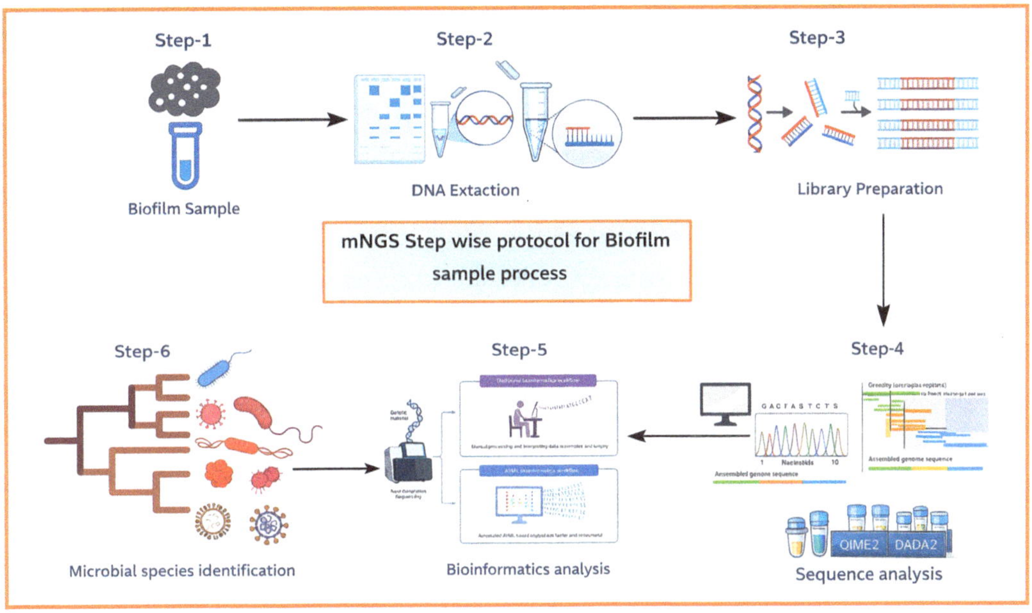

Fig. 1 Stepwise workflow for metagenomic next-generation sequencing (mNGS) of biofilm samples. Schematic representation of the biofilm sample processing pipeline for microbial community characterization

conditions, and handling history should be recorded, as biofilm composition and tightly adherent cells can impact downstream nucleic acid extraction and analysis of biofilm-specific genes, including AMR and virulence factors (Fig. 1).

3.2.1 Biofilm Sampling

- Medical devices or implants—catheters, prosthetic surfaces: Aseptically remove the sample containing the device and place it in a sterile container. If possible, use sonication or vortexing in sterile buffer to detach biofilm cells from the surface.
- Tissue-associated biofilms—Use sterile swabs, scraping instruments, or biopsy punches. Avoid excessive pressure to prevent destruction of the extracellular matrix, which may contain extracellular DNA (eDNA) and secreted biofilm proteins.
- Mixed or heterogeneous samples—Consider multiple sampling points to capture spatial variation in the biofilm [6].

3.2.2 Stabilization of the Sample

- Transfer samples into a nucleic acid stabilization solution (commercial RNA/DNA preservative) to prevent degradation.
- Alternatively, maintain the sample on ice for immediate processing. Rapid freezing at −80 °C is recommended if the sample cannot be processed immediately.

3.2.3 Sample Handling Precautions

- Use sterile, nuclease-free instruments and containers to avoid exogenous DNA/RNA contamination. Minimize sample handling time.
- Limit repeated freeze-thaw cycles to preserve RNA integrity for metatranscriptomics studies.

3.2.4 Meta Documentation

- Source: device type, tissue location, or anatomical site.
- Date/time of collection to correlate with clinical metadata.
- Storage conditions (e.g., on ice, stabilization buffer, −80 °C freezer).
- Pre-treatment history (antibiotics, antiseptics, or mechanical cleaning).
- Biofilm maturity/stage (early, mature, dispersal) if assessable, as it may influence microbial composition.
- Environmental conditions during sampling (e.g., aerobic vs anaerobic, temperature).

3.2.5 Quality Control Checkpoints

- Confirm rapid stabilization to prevent RNA degradation, particularly important for downstream metatranscriptomics sequencing.
- Evaluate sample integrity visually (e.g., intact biofilm vs smeared/disrupted material). Log any deviations in handling or storage to interpret downstream sequencing variability.

3.3 Nucleic Acid Extraction Strategy

- To maximize the recovery of microbial DNA and RNA from mixed bacterial and fungal biofilms while maintaining the representativeness of the community.
- This is essential for accurate downstream sequencing analyses, including metagenomics and metatranscriptomics, which inform studies on antimicrobial resistance (AMR) and virulence factors.

3.3.1 Cell Lysis Optimization

- Employ a combination of mechanical, chemical, and enzymatic methods to achieve comprehensive cell lysis [7].
- Mechanical Lysis: Use bead-beating or sonication to disrupt cell walls, especially effective for Gram-positive bacterial and fungal cells.
- Chemical Lysis: Apply detergents such as SDS or Triton X-100 to solubilize cell membranes.
- Enzymatic Lysis: Incorporate enzymes like lysozyme to degrade peptidoglycan layers in bacterial cell walls and chitinase for fungal cell walls.

- These combined approaches enhance the efficiency of nucleic acid extraction from complex biofilm matrices (*see* **Note 1**).

3.3.2 Extraction Kit Selection

- Choose extraction kits validated for biofilm samples to ensure high yield and purity of nucleic acids.
- The DNeasy Blood and Tissue Kit demonstrated the highest efficiency, yielding significantly more total dsDNA in samples from both healthy and periodontitis-affected sites [8].
- The Power Water-modified kit exhibited less preferential bias regarding low abundance OTUs and better captured overall bacterial diversity compared to the DNeasy PowerSoil kit [9].

3.3.3 Incorporation of Extraction Blanks and Negative Controls

- Process extraction blanks and negative controls (no template DNA/reagent) in parallel to monitor contamination.
- This practice is essential to detect any contamination introduced during sampling, processing, or DNA extraction that may contribute to background detection of microbes and antimicrobial resistance genes (ARGs) [10].
- Regular inclusion of these controls is recommended as a standard practice in metagenomic investigations (*see* **Note 2**).

3.3.4 Assessment of Nucleic Acid Quality and Integrity

- Evaluate the quality and integrity of extracted nucleic acids using spectrophotometric methods (Nanodrop) and fluorometric assays (Qubit).
- Perform agarose gel electrophoresis to assess the integrity of DNA and RNA.
- These assessments ensure that the nucleic acids are of sufficient quality for downstream sequencing applications.

3.4 Sequencing Strategy and Analytical Framework

Goal To select appropriate sequencing platforms and library preparation strategies that align with the study's objectives, ensuring comprehensive analysis of microbial communities and their functional potentials.

3.4.1 Amplicon Sequencing (16S rRNA/ITS)

- *Purpose:* Provides sensitive and cost-effective profiling of bacterial and fungal community composition.
- *Method:* Targets conserved regions (V3-V4 for 16S rRNA) to amplify and sequence specific taxonomic markers.
- *Limitations:*
  - Limited functional insight; does not provide information on gene content or metabolic pathways.

- Potential primer bias and incomplete reference databases may affect species-level identification.
- *Applications:* Suitable for profiling microbial diversity and identifying dominant taxa in a sample.

3.5 Shotgun Metagenomics

- *Purpose:* Enables hypothesis-free identification of all detectable taxa (bacteria, fungi, viruses) and simultaneous detection of functional genes, including those conferring antimicrobial resistance or biofilm-associated virulence.
- *Method:* Randomly fragments total DNA, followed by sequencing and bioinformatics analysis to reconstruct genomes and annotate functional genes.
- *Requirements:*
  - Deeper sequencing coverage to achieve sufficient depth for rare taxa and genes.
  - Rigorous host DNA depletion to minimize background noise and enhance microbial signal.
- *Applications:* Ideal for comprehensive microbial community analysis, functional gene profiling, and detection of novel or unculturable organisms.

3.6 Metatranscriptomics

- *Purpose:* Reveals transcriptionally active members and their expressed pathways, offering insights into metabolic and regulatory activity within biofilms.
- *Challenges:*
  - RNA is inherently unstable; its half-life is in the range of minutes.
  - Low microbial biomass in clinical samples can result in low RNA yields.
  - Contamination with host RNA can complicate data interpretation.
  - Requires specialized protocols for RNA stabilization, enrichment, and purification.
- *Applications:* Useful for studying gene expression profiles, identifying active metabolic pathways, and understanding microbial responses to environmental changes or antimicrobial treatments.

3.7 Sequencing Depth and Platform Selection

3.7.1 Sequencing Depth Considerations

- *Taxonomic Profiling:* A sequencing depth of approximately one million reads per sample is generally sufficient to achieve less than 1% dissimilarity to the full taxonomic composition.
- *Antimicrobial Resistance (AMR) Gene Detection:* At least 80 million reads per sample are required to recover the full richness of

different AMR gene families present in the sample. Additional allelic diversity of AMR genes may still be discovered at 200 million reads per sample [11].

- *Strain-Level Resolution:* Achieving strain-level resolution requires ultra-deep sequencing. A shallow sequencing depth may lead to missing important information and result in biased results. The sufficient sequencing depth for strain-level analysis remains unclear, but deeper sequencing increases the ability to detect low-abundance organisms and resolve strain-level heterogeneity [12].
- *Host DNA Depletion:* In clinical samples with high host DNA content, host DNA depletion methods can increase microbial read counts, species richness, and predicted functional richness. The increase in species and predicted functional richness is mediated by higher effective sequencing depth [13].

3.8 Sequencing Platform Selection

3.8.1 Illumina Short-Read Sequencing

- *Advantages:* High accuracy, high throughput, and cost-effective for large-scale studies.
- *Limitations:* Short read lengths (~150–300 bp) may hinder assembly of complex genomes and detection of structural variants [14].

3.8.2 Oxford Nanopore Technologies (ONT)

- *Advantages:* Long read lengths (up to 2 Mb), real-time sequencing, and portability.
- *Limitations:* Higher error rates compared to short-read platforms; requires more computational resources for data analysis.

3.8.3 PacBio (SMRT) Sequencing

- *Advantages:* Long read lengths (up to 100 kb), high consensus accuracy, and effective for resolving complex genomic regions.
- *Limitations:* Higher cost per base compared to Illumina; lower throughput.

3.9 Pilot Studies and Benchmarking

- *Pilot Datasets:* Conduct pilot sequencing runs to assess the adequacy of chosen sequencing depth and platform for the specific sample type and research objectives.
- *Literature Benchmarks:* Review recent studies with similar sample types and research goals to inform decisions on sequencing depth and platform selection.

3.10 Bioinformatics Workflow and Analytical Modules

3.10.1 Data Processing and Bioinformatics Pipeline

To implement a modular and scalable bioinformatics pipeline that ensures comprehensive analysis of metagenomic data, from quality control to functional annotation, while maintaining flexibility to accommodate various study objectives.

3.10.2 Quality Control

- Adapter Trimming and Quality Filtering: Utilize tools like Trim Galore, Cutadapt, or Fastp to remove adapter sequences and low-quality bases, ensuring high-quality reads for downstream analysis [15].
- Read Length and Complexity Filtering: Employ FASTX-Toolkit or BBMap to filter out reads that are too short or have low complexity, which may introduce noise into the analysis.
- Summary Reporting: Generate comprehensive quality reports using FastQC or MultiQC to assess read quality, GC content, and other metrics, facilitating informed decision-making for subsequent steps.

3.10.3 Host Read Removal

- Host Genome Subtraction: Apply tools like KneadData, DeconSeq, or CLEAN to remove host DNA sequences, enriching the microbial content and improving the sensitivity of microbial detection [16].
- Reference Databases: Utilize comprehensive host genome references, such as the human genome (hg38), to ensure effective subtraction of host sequences.

3.10.4 Taxonomic Profiling

- Marker Gene-Based Classification: Employ tools like MetaPhlAn 4 or Qmatey for taxonomic profiling based on marker genes, enabling accurate identification of microbial taxa [17].
- K-mer-Based Algorithms: Utilize k-mer-based classifiers like Kraken2 or Centrifuge for rapid and sensitive taxonomic assignment, especially in large datasets.
- Reference Databases: Reference curated databases such as RefSeq, GTDB for bacteria, and UNITE for fungi to ensure comprehensive taxonomic coverage.

3.10.5 Functional Annotation

- Gene-Level Analysis: Use tools like Prokka, MetaGene Mark, or RFW for gene prediction and functional annotation, enabling identification of genes related to antimicrobial resistance (AMR), virulence factors, and biofilm-associated functions [18].
- Functional Databases: Reference curated functional databases such as CARD for AMR genes, VFDB for virulence factors, and KEGG or COG for metabolic pathways to ensure comprehensive functional annotation.

3.10.6 Contamination Control

- Negative and Positive Controls: Incorporate negative (no-template) and positive (mock community) controls in the sequencing process to monitor for contamination and assess the accuracy of the analysis.

- Evaluation Metrics: Assess contamination by comparing the presence of expected taxa and functional genes in controls versus experimental samples, ensuring analytical reliability [19].

3.10.7 Modular and Scalable Framework

- Pipeline Integration: Integrate individual modules into a cohesive pipeline using workflow managers like Snakemake, Nextflow, or Galaxy, facilitating reproducibility and scalability.
- Resource Allocation: Optimize computational resource allocation based on dataset size and study objectives, ensuring efficient processing of large-scale metagenomic datasets.
- Benchmarking: Regularly benchmark the pipeline's performance using mock communities or simulated datasets to assess accuracy and sensitivity [13].

3.11 Procedure for Metagenomic Sequencing and Data Analysis

3.11.1 Clinical Sample Collection and Processing

1. Biofilm specimens were collected aseptically from the infection site using needle aspiration, swabbing, curettage, or surgical debridement, depending on clinical accessibility. When feasible, ultrasound or image-guided aspiration was performed to ensure accurate sampling and to minimize contamination.
2. Immediately following collection, each specimen was divided into aliquots for different analyses. For conventional microbiological testing, a portion of the sample was inoculated into both aerobic and anaerobic culture bottles (e.g., BD Biosciences, Sparks, USA).
3. A second aliquot designated for metagenomic next-generation sequencing (mNGS) was placed in a sterile, nuclease-free container without any additives.
4. All specimens were labeled and processed under aseptic conditions. Culture samples were transported to the microbiology laboratory within 1 hour of collection.
5. Specimens intended for mNGS analysis were immediately frozen at −20 °C to preserve nucleic acid integrity and subsequently shipped on dry ice to a certified molecular diagnostics facility within 24 hours for sequencing preparation and analysis.
6. For intraoperative cases, tissue samples were obtained from areas showing the most pronounced inflammatory or biofilm-related changes. These tissues were divided and processed following the same protocol as fluid samples for both culture and molecular testing [13].

3.11.2 Sample Processing and Metagenomic Sequencing

1. Immediately after collection, approximately 2 mL of each biofilm specimen was placed into a sterile microcentrifuge tube and heat-inactivated at 80 °C for 10 min to maintain biosafety.

2. Mechanical disruption of the biofilm matrix was then performed to release microbial cells. Briefly, 0.8 mL of the inactivated sample was transferred into a 1.5 mL microcentrifuge tube containing approximately 2 g of sterile 0.5 mm glass beads.
3. The mixture was agitated on a bead-beating platform at 3000 rpm for 15 min to achieve efficient homogenization.
4. After homogenization, the samples were centrifuged at 13,000 rpm for 10 min to pellet cell debris.
5. A supernatant volume of approximately 600 μL was transferred to a clean 2.0 mL microcentrifuge tube for DNA extraction.
6. Total DNA was extracted using a commercial microbial DNA extraction kit (e.g., TIANamp Micro DNA Kit, Tiangen Biotech, China, or equivalent) following the manufacturer's protocol.
7. The quality and quantity of extracted DNA were assessed using a Qubit fluorometer (Thermo Fisher Scientific, USA) and agarose gel electrophoresis.
8. Sequencing libraries were prepared according to standard protocols for short-read metagenomic sequencing.
9. Briefly, extracted DNA was subjected to random fragmentation, followed by end-repair, adapter ligation, and PCR amplification to generate double-stranded DNA libraries.
10. Library quality and insert size distribution were evaluated using an Agilent 2100 Bioanalyzer (Agilent Technologies, USA) and quantified by Qubit fluorometry.
11. Qualified libraries were denatured and circularized to generate single-stranded circular DNA molecules, which were amplified via rolling circle amplification (RCA) to form DNA nanoballs (DNBs).
12. The DNBs were subsequently loaded onto flow cells for high-throughput sequencing using a BGISEQ (or equivalent) platform with 50 bp single-end reads [20–22].

3.11.3 Library Preparation for Biofilm-Derived mNGS

Following DNA extraction from biofilm specimens, library construction was performed according to the following workflow:

1. The extracted genomic DNA (minimum input ~100 ng, or as low as ~10–20 ng for low-biomass biofilm samples) was first fragmented to an average insert size of ~300 bp using a focused-ultrasonication or enzymatic shearing kit.
2. Fragmented DNA was then subjected to end-repair and A-tailing to generate blunt-ended, 3′-adenylated molecules.

3. Next, sequencing adapters containing sample indices/barcodes were ligated to the DNA fragments to enable multiplexed sequencing.
4. The adapter-ligated DNA was purified (e.g., magnetic-bead cleanup) to remove unlighted adapters and fragments outside the desired size range.
5. A minimal number of PCR amplification cycles (e.g., 6–-10 cycles) were applied to enrich for adapter-ligated molecules, while avoiding amplification bias—particularly important for the complex, low-biomass biofilm samples. (See Busi et al. 2020 for optimisation of low-biomass metagenomic libraries.) [23].
6. Library quality and concentration were assessed by an electrophoretic method (e.g., Agilent 2100 Bioanalyzer) and by fluorescence quantitation (e.g., Qubit).
7. Qualified libraries were normalized, pooled equimolarly, denatured if required, and loaded onto the sequencing platform for high-throughput shotgun sequencing (e.g., 50 bp single-end or 150 bp paired-end reads) as per manufacturer recommendations.
8. Throughout the protocol, care was taken to minimize contamination, avoid adapter-dimer formation, and maintain representation of low-abundance taxa—critical for biofilm-derived communities where fastidious or uncultivable organisms may be present [24].

3.11.4 Bioinformatic Analysis

1. Raw sequencing reads underwent a standardized bioinformatic processing pipeline to ensure data quality and accurate microbial identification (Table 1).
2. Initial quality control was performed using *fastp* to remove adapter sequences, low-quality reads, and short fragments (<35 bp).
3. Reads originating from the human host were computationally filtered out by mapping to the human reference genome (hg38) with the *STAR* aligner.
4. Low-complexity and duplicate sequences were then excluded using *PRINSEQ* to minimize background noise.
5. The remaining high-quality, non-host reads were taxonomically classified with *Kraken2* based on a comprehensive pathogen reference database compiled from multiple public resources, including FDA-ARGOS, BV-BRC, EuPathDB, and NCBI GenBank (*see* **Note 3**).
6. The final dataset included bacterial, viral, fungal, and parasitic taxa, and quantitative metrics species-specific read number (SSRN), reads per million (RPM), and genome coverage were

Table 1
Bioinformatic analysis workflow for mNGS data processing

Step	Software /tool	Version /database	Purpose/description	Key parameters
Quality control	*fastp*	v0.20.1	Removal of low-quality reads and short reads (< 35 bp) to ensure high-quality data for downstream analysis	Length cutoff: 35 bp; default quality filters applied
Host sequence removal	*STAR aligner*	v2.7; human reference genome hg38	Computational subtraction of host-derived reads by mapping to the human genome	Default alignment parameters
Low complexity and duplicate removal	*PRINSEQ*	v0.20.4	Elimination of low-complexity and duplicate reads to reduce background noise and bias	Default parameters
Taxonomic classification	*Kraken2*	v2.1.2	Taxonomic classification of remaining reads based on k-mer matching against a curated reference database	Confidence threshold = 0.1
Reference database composition	–	FDA-ARGOS, BV-BRC (v3.28.9), EuPathDB (release 52), NCBI GenBank (release 242)	Comprehensive pathogen reference database integrating bacterial, viral, fungal, and parasitic genomes	Includes 28,516 bacterial taxa, 8046 viral taxa, 2076 fungal taxa, and 429 parasite taxa
Quantitative metrics	–	–	Species-specific read number (SSRN), reads per million (RPM), and genome coverage (%) were calculated for pathogen identification and abundance estimation	–

used to evaluate pathogen abundance and detection confidence. Details of the computational workflow and reference databases are summarized in Table 1 [25].

3.11.5 Pathogen Identification Criteria for Biofilm-Associated Infections

1. Biofilm-derived specimens, microbial reads classified by *Kraken2* were analyzed with adjusted thresholds to account for the typically low microbial load and polymicrobial nature of these infections. Both quantitative and contextual parameters were integrated to differentiate true pathogens from background or contaminant taxa [26].

2. Microbial taxa with a species-specific read number (SSRN) ≥5 or reads per million (RPM) ≥0.5 were considered for further evaluation. For viral species, detection of at least 3 unique reads was accepted due to their generally low abundance in biofilm environments.
3. Organisms representing >5% of total microbial reads or showing consistent detection across technical replicates were prioritized. When multiple microorganisms were detected, relative abundance and clinical plausibility were jointly evaluated to identify dominant pathogens within polymicrobial communities.
4. Organisms exhibiting partial genome coverage (>1%) with high-quality alignments and without evidence of random or spurious mapping were considered more likely to represent genuine colonizers or pathogens.
5. Taxa with read counts at least fivefold higher than those in negative extraction controls or environmental blanks were retained for interpretation.
6. Common environmental or reagent contaminants (e.g., *Ralstonia*, *Pseudomonas fluorescens*, *Cutibacterium acnes*) were excluded unless corroborated by clinical and quantitative evidence.
7. The clinical context, sampling site, and intraoperative findings were integrated with sequencing data to determine the pathogenic relevance of each detected organism. In cases of polymicrobial detection, species with consistent representation across different sample types (e.g., tissue and fluid) were given higher diagnostic weight.
8. Only microorganisms meeting these combined quantitative, bioinformatic, and clinical criteria were reported as probable pathogens associated with biofilm formation [27].

This mNGS-based workflow offers a broad and sensitive strategy for identifying microbial DNA in biofilm-related infections. Through the integration of effective biofilm disruption, refined nucleic acid extraction, deep sequencing, and rigorous bioinformatic analysis, the method enables unbiased profiling of a wide range of microorganisms, including slow-growing and non-culturable species. When combined with conventional microbiological findings and clinical information, this approach provides a reliable framework for characterizing microbial communities and clarifying the etiological agents involved in biofilm-associated diseases.

4 Notes

1. Host DNA Contamination: In clinical samples, abundant host DNA (e.g., human) can overwhelm microbial sequences, reducing the effective sequencing depth for microbial DNA. Strategies such as host DNA depletion or selective microbial DNA amplification are often necessary.
2. Extraction Efficiency: Biofilms are notoriously difficult to lyse due to the protective EPS matrix. Proper mechanical (e.g., bead-beating) and chemical lysis methods must be used to release microbial DNA effectively.
3. Complex Data Analysis: Bioinformatics pipelines are needed to process vast amounts of sequencing data. Specialized tools (e.g., Kraken2, MetaPhlAn, HUMAnN) are used for taxonomic classification and functional annotation.

Competing Interest

The author confirms that there are no financial competing interests associated with this chapter.

References

1. Nam NN et al (2023) Metagenomics: an effective approach for exploring microbial communities and detecting pathogens. Microorganisms 11(5):1124. https://doi.org/10.3390/foods12112140
2. Blanco-Míguez A et al (2023) Extending and improving metagenomic taxonomic profiling: MetaPhlAn4. Nat Biotechnol 41:1–9. https://doi.org/10.1038/s41587-023-01688-w
3. Kim M et al (2024) Host DNA depletion on frozen human respiratory samples improves metagenomic pathogen detection. Microbiome (PMC) 12:115. https://doi.org/10.21203/rs.3.rs-3638876/v1
4. Zhang Y et al (2024) Evaluation of DNA extraction methods for microbial community profiling. Microbiol Spectrum (PMC) 12: e02456–23. https://doi.org/10.1002/mbo3.70007
5. De Almeida OGG et al (2022) In silico metatranscriptomics approach for tracking biofilm-related effectors. Front Microbiol 13:928480. https://doi.org/10.3389/fmicb.2022.928480
6. Kim C-W et al (2025) Benefits and challenges of host depletion methods in clinical metagenomic sequencing. NPJ Biofilms Microbiomes/Nat Microbiol (preview). https://doi.org/10.1038/s41522-025-00762-2
7. Lee J, Chua B, Son A (2024) Recent advances in bacterial lysis techniques for environmental monitoring: a review. Microchem J 207: 111865., ISSN 0026-265X. https://doi.org/10.1016/j.microc.2024.111865
8. Wäge-Recchioni et al (2025) Comparative pilot study of three commercial kits for bacterial DNA extraction from human subgingival biofilm samples collected with a single paper point. J Oral Microbiol 17. https://doi.org/10.1080/20002297.2025.2549035
9. Putri Ratna et al (2024) Enhancing the DNA yield intended for microbial sequencing from a low-biomass chlorinated drinking water. Front Microbiol 15. https://doi.org/10.3389/fmicb.2024.1339844
10. Weyrich LS et al (2019) Laboratory contamination over time during low-biomass sample analysis. Mol Ecol Resour 19(4):982–996. https://doi.org/10.1111/1755-0998.13011
11. Gweon HS et al (2019) The impact of sequencing depth on the inferred taxonomic composition and AMR gene content of metagenomic samples. Environ Microbiome 14(1):7.

https://doi.org/10.1186/s40793-019-0347-1

12. Liu Pu et al (2022) Towards strain-level complexity: sequencing depth required for comprehensive single-nucleotide polymorphism analysis of the human gut microbiome. Frontiers in Microbiology. 13:1664-302X. https://doi.org/10.3389/fmicb.2022.828254
13. Kim M et al (2024) Host DNA depletion on frozen human respiratory samples enables successful metagenomic sequencing for microbiome studies. Res Sq [Preprint] 23:rs.3.rs-3638876. doi:10.21203/rs.3.rs-3638876/v1. Update in: Commun Biol. 2024 Nov 28;7 (1):1590. https://doi.org/10.21203/rs.3.rs-3638876/v1
14. Sajib MSI, et al. Advances in host depletion and pathogen enrichment for clinical metagenomics. Clin Microbiol Rev 2024;37(2): e00012–24. https://doi.org/10.1128/CMR.00012-24. https://doi.org/10.1016/j.jmoldx.2024.05.008
15. Chen YC et al (2025) Optimization of metagenomic next-generation sequencing for pathogen detection in clinical samples. Front Microbiol (PMC). https://doi.org/10.1007/s40291-025-00797-3
16. Gao Y et al (2025) Benchmarking short-read metagenomics tools for removing host contamination. Gigascience 14:giaf004. https://doi.org/10.1093/gigascience/giaf004
17. Blanco-Míguez A, Beghini F, Cumbo F et al (2023) Extending and improving metagenomic taxonomic profiling with uncharacterized species using MetaPhlAn 4. Nat Biotechnol 41:1633–1644. https://doi.org/10.1038/s41587-023-01688-w
18. Mi K et al (2024) RFW captures species-level metagenomic functions by integrating genome annotation information. Cell Report Method 4(12):100932., ISSN 2667-2375. https://doi.org/10.1016/j.crmeth.2024.100932
19. Beghini F et al (2021) Integrating taxonomic, functional, and strain-level profiling: MetaPhlAn 3 and HUMAnN 3. Nat Commun 12:1–11. https://doi.org/10.7554/eLife.65088
20. Kool J et al (2023) Reducing bias in microbiome research: comparative impact of extraction and library strategies. Front Microbiol 14: 109480. https://doi.org/10.3389/fmicb.2023.1094800
21. Wood DE, Lu J, Langmead B (2019) Improved metagenomic analysis with kraken 2. Genome Biol 20:257. https://doi.org/10.1186/s13059-019-1891-0
22. Van Den Bossche T et al (2025) The microbiologist's guide to metaproteomics. Nat Rev Microbiol. https://doi.org/10.1002/imt2.70031. (early access)
23. Busi SB et al (2020) Optimised biomolecular extraction for metagenomic analysis of microbial biofilms from high-mountain streams. Peer J 8:e9973. https://doi.org/10.7717/peerj.9973
24. Shi Y et al (2022) Metagenomic sequencing for microbial DNA in human samples: emerging technological advances. Int J Mol Sci 23:2181. https://doi.org/10.3390/ijms23042181
25. Alcock BP et al (2020) CARD: Antibiotic resistome surveillance with the comprehensive antibiotic resistance database. Nucleic Acids Res 48 (D1):D517–D525. https://doi.org/10.1093/nar/gkz935
26. Liu B (2019) VFDB: a comparative pathogenomic platform for virulence factor research. Nucleic Acids Res 47(D1):D687–D692. https://doi.org/10.1093/nar/gky1080
27. Atriwal T (2021) Mechanistic understanding of Candida albicans biofilm formation. Front Microbiol 12:638609. https://doi.org/10.3389/fmicb.2021.638609

Chapter 13

Transcriptomics (RNA-seq): Exploring and Revealing Gene Expression Patterns During Biofilm Formation

Rajesh Palanisamy

Abstract

RNA-sequencing (RNA-seq) has emerged as the gold standard for comprehensive transcriptomic profiling, enabling unbiased, high-resolution quantification of genome-wide gene expression patterns. This chapter provides a practical guide to designing and executing RNA-seq experiments optimized for exploring and revealing gene expression dynamics during biofilm formation, a complex developmental process characterized by coordinated bacterial cell differentiation and extracellular matrix production. We detail protocols for sample collection, total RNA extraction with quality assessment, library preparation using stranded chemistry, and bioinformatics workflows including quality control, read alignment, and differential expression analysis using DESeq2. Special attention is given to strategies for capturing stage-specific biofilm gene expression, normalization approaches for complex bacterial communities, and validation of results via RT-qPCR. Real-world examples from diverse biofilm formation models illustrate how RNA-seq reveals virulence factor production, quorum sensing-regulated pathways, stress response mechanisms, and antimicrobial resistance gene activation. This chapter equips researchers in microbiology and biofilm biology with practical knowledge and troubleshooting strategies for producing robust, reproducible transcriptomic datasets that advance understanding biofilm pathogenesis.

Key words RNA-seq, Transcriptomics, Biofilm formation, Gene expression, Differential expression, DESeq2, Quorum sensing, Virulence factors, Library preparation, RNA quality control, Bioinformatics

1 Introduction

RNA-sequencing has revolutionized molecular biology by providing unbiased, genome-wide quantification of mRNA transcripts. Unlike microarray-based methods, RNA-seq is not limited to known sequences, enables detection of rare transcripts at single-molecule resolution, and delivers digital readout across an essentially unlimited dynamic range [1, 2]. This technological advance has transformed our ability to study complex biological phenomena such as biofilm formation—a multicellular developmental process fundamental to bacterial survival, persistence, and virulence.

Abirami K (ed.), *Microbial Biofilm Research: Comprehensive Protocols and Best Practices*, Springer Protocols Handbooks, https://doi.org/10.1007/978-1-0716-5535-1_13,

Biofilms represent highly organized bacterial communities embedded in extracellular polymeric substances (EPS), enabling coordinated gene expression that confers antimicrobial resistance, nutrient acquisition, and immune evasion [3, 4]. Unlike planktonic (free-floating) bacteria, biofilm cells exhibit dramatically altered transcriptional programs reflecting physiological adaptation to nutrient-limited microenvironments, differential oxygen availability, and physical crowding. RNA-seq reveals these stage-specific transcriptional changes in unprecedented detail, uncovering hidden molecular programs orchestrating biofilm architecture, matrix production, virulence factor synthesis, and persistence mechanisms [5].

Experimental design is paramount [6]. The biological question dictates whether bulk RNA-seq (average expression across the mixed population) or single-cell RNA-seq (cell-type-specific expression within biofilms) is appropriate. For biofilm studies, bulk RNA-seq from standardized biofilm communities cultured under defined conditions enables high-confidence identification of differentially expressed genes (DEGs) associated with each developmental stage [7]. Alternatively, spatial transcriptomics or single-cell approaches reveal heterogeneous gene expression within structurally distinct biofilm regions. Technical considerations including RNA integrity, library preparation chemistry, sequencing depth sufficient to capture minority populations, and computational approaches for analyzing mixed microbial communities critically impact reproducibility and statistical power [6].

This chapter synthesizes current best practices, guiding readers through experimental design, sample preparation, sequencing, and computational analysis for biofilm transcriptomics. Special focus is placed on practical troubleshooting and quality control to ensure generation of robust datasets suitable for peer-reviewed publication.

2 Materials

2.1 Sample Collection and RNA Extraction

1. Biofilm-forming bacterial strains cultured under standard biofilm growth conditions (static culture, flow cells, or microfluidic devices).
2. Sterile phosphate-buffered saline (PBS), pH 7.4.
3. TRIzol Reagent (Invitrogen, Thermo Fisher Scientific, Waltham, MA).
4. RNeasy Mini Kit with RNase-Free DNase Set (Qiagen, Germantown, MD).
5. Chloroform, isopropanol, ethanol (analytical reagent grade).
6. RNase-free water, pipette tips, and microfuge tubes.

2.2 RNA Quality Assessment

1. Agilent 2100 Bioanalyzer with RNA 6000 Nano Kit.
2. Agilent TapeStation 4200.
3. Qubit 4 Fluorometer with RNA High Sensitivity (HS) Assay Kit (Thermo Fisher).
4. NanoDrop ND-1000 Spectrophotometer.

2.3 Library Preparation

1. Illumina Stranded Total RNA Prep kit with Ribo-Zero Plus (Illumina, San Diego, CA).
2. NEBNext Ultra II Directional RNA Library Prep Kit (New England Biolabs, Ipswich, MA).
3. NEBNext Multiplex Oligos (Index Primers Sets 1–4; NEB).
4. AMPure XP magnetic beads (Beckman Coulter, Indianapolis, IN).
5. Ethanol (molecular biology grade).
6. Nuclease-free water.
7. Magnetic stand for bead separation.

2.4 Sequencing and Bioinformatics

1. Illumina NextSeq 2000 or NovaSeq 6000 sequencer.
2. PhiX control library (Illumina).
3. High-performance computing cluster or local workstation.
4. Bioinformatics software: FastQC (v0.11.9+), Trimmomatic (v0.39), STAR aligner (v2.7.9+), featureCounts (v2.0+).
5. R/RStudio (v4.0+) with packages: DESeq2, ggplot2, pheatmap.
6. Reference genome: biofilm-forming bacterial species reference genome (NCBI or specialty databases).

2.5 RT-qPCR Validation

1. qPCR instrument (Applied Biosystems StepOnePlus or similar).
2. SYBR Green or TaqMan Master Mix (Applied Biosystems).
3. cDNA synthesized from total RNA via reverse transcriptase.
4. Gene-specific oligonucleotide primers spanning introns (see Table 1).

3 Methods

3.1 Experimental Design and Sample Preparation

1. Define biological question: What are the gene expression signatures associated with early, intermediate, and mature biofilm stages? Which genes regulate biofilm development, matrix synthesis, virulence factor production, or antimicrobial resistance? Will you measure time course (6 h, 12 h, 24 h, 48 h) or compare planktonic versus biofilm cells?

Table 1
Representative RT-qPCR primers for validation of biofilm-associated genes

Gene	Gene category	Forward primer (5′–3′)	Reverse primer (5′–3′)	Tm (°C)	Amplicon (bp)	Expected FC
pelA (EPS biosynthesis)	Biofilm development	TGTAGACCATGTAGTTGAGGTCA	AGGTCGGTGTGAACGGATTTG	60	187	↑↑
lasR (quorum sensing)	Virulence regulation	CGATGACATCGACGTGCTGA	TGACGATCACGTCGATCTCA	58	142	↑
rhlR (quorum sensing)	Virulence regulation	AGCATCGGTGTGGTCATGCA	TCGATCGATCGATCGAGCCA	60	165	↑
algD (alginate biosynthesis)	EPS production	CCATCTTCCTGCCCTCAAGA	CTGGGCTCAGACTGTCGATG	60	152	↑
oprI (outer membrane)	Cell envelope	TGTAGACCATGTAGTTGAGGTCA	AGGTCGGTGTGAACGGATTTG	60	179	↑
rpoD (housekeeping control)	Reference gene	ATGACCATGACCACGCTGAA	GCGGTGATCATCTGCGTCTG	60	156	=

2. Design statistical framework: Specify minimum number of biological replicates. For moderate effect sizes comparing biofilm stages or biofilm versus planktonic cells, 4–6 replicates per condition are recommended [6]. Randomize sample collection order and processing steps to minimize batch effects. Plan adequate sequencing depth to detect minority transcripts within biofilm communities.
3. For standard biofilm culture assays: Use defined biofilm growth models appropriate to research questions. Static culture methods (polystyrene microtiter plates or cover slips) enable simple time-course sampling [8]. Flow cell systems provide physiologically relevant shear stress and nutrient gradients. Microfluidic devices allow precise control of microenvironment. Culture biofilms under identical conditions to standardization specifications (temperature, media composition, aeration) across all replicates. Record precise incubation times and environmental parameters.
4. At designated biofilm development stages (early, intermediate, mature), carefully collect biofilm biomass. For static cultures: aspirate culture medium, rinse biofilms 2× in ice-cold PBS to remove planktonic cells, use sterile cotton swabs or cell scrapers to collect biofilm. For flow cells: stop flow, rinse with cold PBS, collect biofilm material. For comparison studies: simultaneously collect planktonic cells from culture supernatant (centrifuge, resuspend pellet in PBS) to enable direct gene expression comparison.
5. Quantify biofilm biomass gravimetrically (dry weight) or via crystal violet staining to document biofilm development stage and consistency across replicates.
6. Immediately lyse biofilm cells in TRIzol (1 mL per 5–10 × 10^6 cells estimated). Mix thoroughly by pipetting up and down 10 times. Incubate at room temperature for 5 min.

3.2 RNA Extraction and Quantification

1. Phase-separate by adding chloroform (0.2 mL per mL TRIzol). Shake vigorously for 15 seconds, incubate 2–3 min at room temperature, then centrifuge (12,000× *g*, 15 min, 4 °C).
2. Transfer aqueous (upper) phase to fresh microcentrifuge tube. Add equal volume of isopropanol, mix gently, incubate 10 min at room temperature.
3. Centrifuge (12,000× g, 10 min, 4 °C) to precipitate RNA as white pellet. Discard supernatant.
4. Wash pellet with 75% ethanol (prepared fresh in RNase-free water), centrifuge (7500× *g*, 5 min, 4 °C). Repeat wash step.

5. Air-dry pellet 5–10 min (do not over-dry; this reduces dissolution). Resuspend in 30–50 μL RNase-free water by incubation at 55–60 °C for 10–15 min with occasional gentle mixing.
6. Quantify total RNA using Qubit HS assay kit per manufacturer's instructions. Target: >1 μg total RNA at ≥50 ng/μL concentration.
7. Assess RNA integrity using Bioanalyzer: RNA Integrity Number (RIN) ≥8 is optimal; RIN ≥7 acceptable for stranded protocols [7]. RIN <7 indicates degradation; consider discarding samples or using rRNA depletion approach (*see* **Note 1**).

3.3 Library Preparation Using Illumina Stranded Protocol

This protocol generates stranded, directional RNA-seq libraries suitable for transcript orientation determination (*see* **Note 2**) [9].

1. Fragment total RNA using fragmentation buffer (94 °C, 4–6 min depending on desired fragment length; target 200–300 bp average size).
2. Synthesize first-strand cDNA: Add random hexamers and reverse transcriptase. Incubate 42 °C for 15 min, then 70 °C for 15 min to inactivate enzyme.
3. Synthesize second-strand cDNA using DNA polymerase I. This step incorporates dUTP, which allows selective uracil-DNA glycosylase treatment in subsequent steps to maintain strand information.
4. End-repair: Incubate cDNA with end-repair enzyme mix to phosphorylate and blunt all 5′ and 3′ ends (37 °C, 30 min).
5. Add A-tailing reaction (Klenow polymerase 3′ → 5′ exo-minus, dATP; 37 °C, 30 min) to add single adenine residue to all 3′ ends.
6. Adapter ligation: Mix cDNA with Illumina paired-end adapters (containing unique index sequences for sample multiplexing) and DNA ligase. Incubate 15 °C for 15 min.
7. Selective uracil-DNA glycosylase digestion: This enzymatic step degrades the dUTP-containing second strand, regenerating the desired first-strand cDNA library.
8. PCR amplification: Perform 12–15 cycles of PCR (optimize cycle number empirically; excessive cycles introduce bias and sequencing artifacts [7]).
9. Post-PCR cleanup: Add AMPure XP magnetic beads (0.8× bead:sample ratio) to remove residual adapters and primer dimers. Mix well, incubate 5 min at room temperature, place on magnetic stand, wash with 80% ethanol twice, air-dry, and elute in 30 μL nuclease-free water (*see* **Note 3**).

10. Verify library quality using Agilent TapeStation: Peak should be 250–350 bp. Quantify final library with Qubit dsDNA HS assay; target ≥2 nM.

3.4 Sequencing

1. Load library at 1–2 pM on Illumina sequencer (NextSeq 2000 or NovaSeq 6000). Specify paired-end mode (75 × 75 bp or 100 × 100 bp).
2. Plan sequencing depth: 20–30 million paired-end reads per sample provide adequate coverage for differential expression studies [7, 10].
3. Include 5–10% PhiX control library in lane to monitor sequencing quality in real-time.
4. Perform bcl2fastq demultiplexing to generate per-sample FASTQ files. Retain both forward and reverse reads (R1 and R2 FASTQ files).

3.5 Bioinformatics Pipeline

3.5.1 Quality Assessment

1. Run FastQC on raw FASTQ files to assess read quality, adapter contamination, and GC content. Evaluate for: Mean quality score ≥ Q30, adapter contamination <5%.
2. Aggregate results using multiQC across all samples (*see* **Note 4**).

3.5.2 Adapter and Quality Trimming

1. Remove Illumina adapters and low-quality bases using Trimmomatic [9]:

```
trimmomatic PE sample_R1.fastq.gz sample_R2.fastq.gz
    sample_R1_paired.fastq.gz sample_R1_unpaired.fastq.
  gz
    sample_R2_paired.fastq.gz sample_R2_unpaired.fastq.
  gz
    ILLUMINACLIP:TruSeq3-PE-2.fa:2:30:10
    SLIDINGWINDOW:4:20 MINLEN:25,
```

2. Inspect trimming statistics. Typically, >95% of reads remained post-trimming.

3.5.3 Genome Alignment

1. Align trimmed reads using STAR two-pass mode [11]:

```
STAR --runThreadN 8
    --genomeDir./star_index
    --readFilesIn sample_R1_paired.fastq.gz sample_R2_-
  paired.fastq.gz
    --readFilesCommand zcat
    --outFileNamePrefix sample_
    --twopassMode Basic
    --outSAMtype BAM SortedByCoordinate
```

2. Quality check: Target ≥80% uniquely mapped reads.

3.5.4 Read Quantification

1. Count reads mapping to known genes using featureCounts [12]:

```
featureCounts -p -s 2
        -a biofilm_species.gtf
        -o counts_matrix.txt
        sample1_Aligned.sortedByCoord.out.bam
        sample2_Aligned.sortedByCoord.out.bam
```

Flag -s 2 specifies reverse-strand library (Illumina Stranded protocol). Flag -p indicates paired-end mode (*see* **Note 5**).

3.5.5 Differential Expression Analysis with DESeq2

1. Prepare matrix and metadata in R:

```
library(DESeq2)
        countData <- read.csv("counts_matrix.txt", row.
    names=1)
        colData <- data.frame(
        biofilm_stage = c
    ("early","early","early","intermediate",
        "intermediate","intermediate","mature","mature"),
    batch = c("Day1","Day1","Day2","Day1",
    "Day2","Day2","Day1","Day2")
        )
        rownames(colData) <- colnames(countData)dds <-
    DESeqDataSetFromMatrix(
        countData = countData,
        colData = colData,
        design = ~ batch + biofilm_stage
        )
```

2. Filter lowly expressed genes (≥10 counts in ≥2 samples) [10]:

```
keep <- rowSums(counts(dds) >= 10) >= 2
        dds <- dds[keep,]
```

3. Perform differential expression analysis:

```
dds <- DESeq(dds)
        res_intermediate_vs_early <- results(dds, contrast=c
    ("biofilm_stage","intermediate","early"), alpha=0.05)
        res_mature_vs_early <- results (dds, contrast=c("bio-
    film_stage","mature","early"), alpha=0.05)
```

The DESeq() function automatically estimates size factors using median-of-ratios normalization [7], estimates gene-wise dispersion, and performs Wald test with adjusted p-values (Benjamini-Hochberg correction).

4. Extract significant DEGs:

```
resSig_int <- res_intermediate_vs_early[which(res_interme-
     diate_vs_earlylog2FoldChange) > 1),]
        resSig_mat <- res_mature_vs_early[which(res_ma-
    ture_vs_earlylog2FoldChange) > 1),]
```

3.6 Visualization and Functional Enrichment

1. Generate MA plot (mean average expression vs. $\log_2$ fold change):

   plotMA(res_mature_vs_early, ylim=c(−5, 5)),

2. Create volcano plot:

   plot(res_mature_vs_early *log2FoldChange*, − *log* 10(*res$_m$ature-$_v$s$_e$arly* padj), main="Volcano Plot: Mature vs Early Biofilm")
   abline(v=c(-1, 1), h=-log10(0.05), col="red", lty=2)

3. Generate heatmap of top 50 DEGs:

   library(pheatmap)
   topGenes <− head(order(res_mature_vs_early $padj), 50)
   pheatmap(log2(counts(dds, normalized = TRUE) [topGenes,]))

4. Perform functional enrichment analysis using tools such as DAVID, Metascape, or ClusterProfiler to identify overrepresented Gene Ontology (GO) terms, KEGG pathways, or custom biofilm-specific gene sets associated with each developmental stage [13].

3.7 RT-qPCR Validation

1. Select 8–12 top DEGs plus 2–3 stable reference genes for validation on independent replicates (*see* **Note 6**) [14].
2. Prepare cDNA via reverse transcription from 1 μg total RNA per sample.
3. Design RT-qPCR primers (18–22 bp, Tm 58–60 °C) spanning introns to avoid genomic DNA amplification [15].
4. Perform RT-qPCR using SYBR Green or TaqMan chemistry [16].
5. Calculate fold changes using ΔΔCt method [17]:

   $\log_2$ FC (qPCR) = $2^{(-\Delta\Delta Ct)}$.

6. Compare RNA-seq $\log_2$FC versus RT-qPCR $\log_2$FC: Pearson correlation coefficient should exceed 0.80 (*see* **Note 7**) [14, 16].

4 Notes

1. *rRNA contamination:* If >80% of reads maps to ribosomal RNA, perform rRNA depletion using Ribo-Zero (Illumina) or Ribo-X kits. This boosts coding-sequence read coverage by three- to tenfold [10].
2. *Batch effects:* Include batch variable in DESeq2 design formula (e.g., design = ~ batch + biofilm stage). This regresses out batch variation while preserving true biological condition effects. Post hoc correction using ComBat or SVA is also effective [18].

3. *Mixed microbial communities:* For biofilms composed of multiple bacterial species, sequencing reveals expression of each species simultaneously if reference genomes are available. Computational deconvolution or separate alignment to species-specific genomes enables species-specific differential expression analysis [19]. Alternatively, sequencing mixed biofilms followed by metagenomics-based sorting enables species-level transcript quantification [20].
4. *Planktonic versus biofilm comparison:* Direct comparison reveals dramatic transcriptional reprogramming. Prepare planktonic cell samples simultaneously with biofilm samples to minimize batch effects. Include planktonic cells from same culture media and timepoints as biofilm samples for rigorous controls [21].
5. *Low-input samples (<100 ng RNA):* Biofilm samples sometimes yield limited RNA, particularly if extracted from biofilm regions with high EPS content relative to cellular material. SMART-seq and SMARTer kits (Clontech) are specifically designed for full-length cDNA synthesis from low-input material while preserving isoform information.
6. *Quality control checklist:*

   ✓ RIN ≥7; median RIN across samples ≥8

   ✓ >80% uniquely mapped reads

   ✓ Correlation between replicates (Pearson $r > 0.95$)

   ✓ >80% of top DEGs validated by RT-qPCR ($r > 0.80$)

   ✓ Functional enrichment results concordant with biofilm biology hypothesis

   ✓ Biofilm biomass quantification confirms developmental stage across replicates
7. *Common pitfalls to avoid:*
   - Over-trimming reduces mapping rates. Use SLIDINGWINDOW: 4:20, MINLEN:25.
   - Ignoring lowly expressed genes inflates multiple-testing burden. Filter appropriately (≥10 counts in ≥2 samples).
   - Inadequate replication. Minimum 4–6 replicates per condition recommended for statistical power.
   - Failure to validate independently. RT-qPCR validation on separate biological samples is essential [16].
   - Misspecifying library chemistry in featureCounts (use -s 0 unstranded, -s 1 forward, -s 2 reverse) reverses expression estimates.
   - Insufficient sequencing depth for minority transcripts within heterogeneous biofilms. Aim for ≥30 million reads per sample.

- Poor biofilm standardization across replicates. Maintain consistent culture conditions, timing, and collection methods.

Competing Interest

The author confirms that there are no financial competing interests associated with this chapter.

References

1. Wang Z, Gerstein M, Snyder M (2009) RNA-seq: a revolutionary tool for transcriptomics. Nat Rev Genet 10:57–63
2. Conesa A, Madrigal P, Tarazona S et al (2016) A survey of best practices for RNA-seq data analysis. Genome Biol 17:13
3. Flemming H-C, Wingender J (2010) The biofilm matrix. Nat Rev Microbiol 8:623–633
4. Hall-Stoodley L, Costerton JW, Stoodley P (2004) Bacterial biofilms: from the natural environment to infectious diseases. Nat Rev Microbiol 2:95–108
5. Toutain CM, Carevic MP, O'Toole GA (2007) Molecular basis of biofilm development by pseudomonads. Microbiol Mol Biol Rev 71: 646–662
6. Zhao S, Zhang B (2015) A comprehensive review of computational methods for small RNA discovery using next-generation sequencing data. Comput Biol Chem 59:121–134
7. Anders S, Huber W (2010) Differential expression analysis for sequence count data. Genome Biol 11:R106
8. O'Toole GA, Kolter R (1998) Initiation of biofilm formation in *Pseudomonas fluorescens* WCS365 proceeds via multiple, convergent signalling pathways. Mol Microbiol 28:449–461
9. Bolger AM, Lohse M, Usadel B (2014) Trimmomatic: a flexible trimmer for Illumina sequence data. Bioinformatics 30:2114–2120
10. Love MI, Huber W, Anders S (2014) Moderated estimation of fold change and dispersion for RNA-seq data with DESeq2. Genome Biol 15:550
11. Dobin A, Davis CA, Schlesinger F et al (2013) STAR: ultrafast universal RNA-seq aligner. Bioinformatics 29:15–21
12. Liao Y, Smyth GK, Shi W (2014) Feature counts: an efficient general-purpose program for assigning sequence reads to genomic features. Bioinformatics 30:923–930
13. Huang DW, Sherman BT, Lempicki RA (2009) Systematic and integrative analysis of large gene lists using DAVID bioinformatics resources. Nat Protoc 4:44–57
14. Hellemans J, Mortier G, De Paepe A, Speleman F (2007) qBase relative quantification framework and software for management and automated analysis of real-time quantitative PCR data. Genome Biol 8:R19
15. Livak KJ, Schmittgen TD (2001) Analysis of relative gene expression data using real-time quantitative PCR and the 2^-ΔΔCt method. Methods 25:402–408
16. Everaert C et al (2017) Benchmarking of RNA-sequencing analysis workflows using whole transcriptome qPCR data and long RNA-seq reads. Sci Rep 7:1559
17. Johnson WE, Li C, Rabinovic A (2007) Adjusting batch effects in microarray expression data using empirical Bayes methods. Biostatistics 8: 118–127
18. Leek JT, Johnson WE, Parker HS, Jaffe AE, Storey JD (2012) The sva package for removing batch effects and other unwanted variation in high-throughput experiments. Bioinformatics 28:882–883
19. Woldringh CL, Mulder E, Huls PG, Cohen JA (1977) Abrupt transition to smaller cells and decreased surface area of the nucleoid upon starvation of Escherichia coli. J Bacteriol 131: 270–279
20. Stewart EJ, Madden R, Paul G, Taddei F (2005) Aging and death in an organism that reproduces by fission. PLoS Biol 3:e295
21. Sauer K (2003) The genomics and proteomics of biofilm formation. Genome Biol 4:219

Part V

Physical and Mechanical Methods

Chapter 14

Real-Time Monitoring of Microbial Biofilms Using Quartz Crystal Microbalance

Sujith P

Abstract

Biofilm formation on solid surfaces is a dynamic, multistep process influenced by microbial physiology, surface chemistry, and hydrodynamic conditions. In this study, an integrated, culture-based, and biophysical approach was employed to investigate biofilm adhesion, growth, and structural properties in real time. Quartz crystal microbalance with dissipation monitoring (QCM-D) was used to monitor biofilm development on gold- and titanium-coated quartz crystals by measuring changes in resonance frequency and energy dissipation, providing insights into biomass accumulation and viscoelastic behavior. A flow cell system was coupled to the QCM chamber to enable continuous monitoring of biofilm formation under controlled shear stress and nutrient flow. Surface-associated biofilms formed on quartz crystals were further characterized using atomic force microscopy (AFM) to examine nanoscale morphology and topographical features. To validate QCM-derived biomass measurements, biofilm formation was independently quantified using crystal violet staining and spectrophotometric analysis. Standardized cleaning and reuse protocols for QCM crystals were implemented to ensure experimental reproducibility. This combined methodology offers a robust and reproducible framework for real-time biofilm analysis, enabling quantitative assessment of biofilm attachment kinetics, mechanical properties, and biomass under static and dynamic conditions. The protocol is broadly applicable to environmental, industrial, and biomedical biofilm studies.

Key words Biofilm formation, Quartz crystal microbalance with dissipation (QCM-D), Atomic force microscopy (AFM), Flow cell system, Biofilm viscoelasticity, Surface adhesion, Crystal violet assay, Culture-based biofilm quantification

1 Introduction

Biofilms are surface-attached microbial communities encased in a self-produced extracellular matrix that confers protection against environmental stressors and antimicrobial agents [1, 2]. Conventional methods for biofilm assessment often provide only endpoint data and fail to capture the dynamic nature of biofilm development.

Quartz crystal microbalance (QCM) is a highly sensitive, label-free technique that monitors real-time changes in mass on a quartz sensor surface by measuring shifts in resonance frequency [3].

Abirami K (ed.), *Microbial Biofilm Research: Comprehensive Protocols and Best Practices*, Springer Protocols Handbooks,
https://doi.org/10.1007/978-1-0716-5535-1_14,

When combined with dissipation monitoring (QCM-D), it also provides insights into the viscoelastic properties of the biofilm, allowing a more comprehensive understanding of its structure and behavior under physiological or flow conditions [4]. These capabilities make QCM an invaluable tool for studying biofilm formation, adhesion kinetics, and response to antimicrobials.

2 Materials

2.1 General Microbiological Materials

- Sterile growth medium (e.g., nutrient broth, tryptic soy broth, or other suitable medium).
- Agar plates for culture maintenance.
- Sterile distilled/deionized water.
- Sterile pipette tips and micropipettes.
- Sterile culture tubes and flasks.
- Glycerol stocks (for long-term microbial storage, if applicable).

2.2 QCM Setup

- Quartz crystal microbalance with dissipation monitoring.
- Gold-coated quartz crystals (10 MHz AT-cut).
- Titanium-coated quartz crystals (10 MHz AT-cut).
- Teflon cover and open cell chamber (~200 μL capacity).
- Moisture-preserving cap or water reservoir (to prevent evaporation during long-term measurements).
- Data acquisition software provided by manufacturer.
- Data analysis software.

2.3 Biofilm Characterization

- Atomic force microscope (AFM).
- AFM cantilevers (e.g., aluminum back-coated, nominal tip radius ~7 nm, spring constant ~26 N/m).
- NanoScope Analysis software.

2.4 Biofilm Quantification Assay

- Crystal Violet (CV) dye (0.5% stock solution prepared with distilled water and methanol).
- 96-well microtiter plates (for parallel control assays).
- ELISA plate reader.

2.5 Cleaning and Reuse of QCM Crystals

- Sulfuric acid (H_2SO_4, concentrated).
- Hydrogen peroxide (H_2O_2, 30%).
- Acetone (analytical grade).
- Isopropanol (analytical grade).
- Milli-Q water (ultrapure).
- Nitrogen gas stream (for drying).
- Glass beakers and steel tweezers (for handling crystals).

3 Methods

3.1 Preparation of Microbial Cultures

1. Microbial cultures of interest were maintained on agar plates and stored as glycerol stocks at −80 °C. For experiments, single colonies were inoculated into sterile liquid growth medium and incubated overnight at the organism's optimal growth temperature under continuous shaking (≈200 rpm).
2. Overnight cultures were adjusted to an appropriate inoculum density for biofilm initiation (e.g., optical density or CFU/mL equivalent) before use [2] (*see* **Note 1**).

3.2 Quartz Crystal Microbalance (QCM) Measurements

1. QCM-D experiments were performed using a quartz crystal-based device. Gold-coated and titanium-coated quartz oscillators (10 MHz, AT-cut) were mounted in an open cell chamber (~200 μL capacity), sealed with a Teflon cover, and enclosed with a moisture-preserving cap to minimize evaporation (*see* **Note 2**).
2. Baseline stability was achieved by loading sterile growth medium until resonance frequency and dissipation signals stabilized (~1 h). The microbial inoculum was then introduced, and measurements were recorded continuously for up to 24 h. Both frequency and dissipation shifts were monitored in real time using the manufacturer's software and analyzed using data analysis tools (e.g., OriginPro). This methodology allows assessment of biofilm adhesion, growth, and viscoelasticity [3, 4].

3.3 Biofilm Characterization Using AFM

1. Following QCM measurements, biofilms formed on quartz crystal surfaces were characterized by atomic force microscopy (AFM). Imaging was carried out in tapping mode using aluminum back-coated cantilevers (nominal tip radius ~7 nm, spring constant ~26 N/m, resonant frequency ~300 kHz).
2. Scan sizes of 10–20 μm were recorded at a resolution of 512 pixels/line, with a scan rate of ~1 line/s. Images were processed and analyzed using NanoScope Analysis software to examine surface morphology and biofilm topography [5] (*see* **Note 3**).

3.4 Integration of Flow Cell System

1. A flow cell system was coupled with the QCM chamber to enable real-time monitoring of biofilm development under dynamic flow conditions. The system consisted of a peristaltic pump connected to inlet and outlet ports of the QCM chamber via sterile silicone tubing (inner diameter: 1.0–1.5 mm). Flow rates were maintained between 0.1 and 1.0 mL/min depending on the experimental design (*see* **Note 4**).
2. The flow system was sterilized prior to each experiment by flushing with 70% ethanol followed by sterile distilled water.

Growth medium or buffer was then circulated through the chamber to equilibrate temperature and establish a stable baseline. After baseline stabilization, the microbial suspension was introduced through the flow inlet for a defined adhesion period (e.g., 30 min–1 h) under static or low-flow conditions. Subsequently, fresh medium was perfused continuously to simulate nutrient exchange and shear stress typical of natural environments [6, 7] (*see* **Note 5**).

3. This configuration enables simultaneous monitoring of frequency and dissipation changes in real time, allowing quantification of biofilm attachment kinetics, viscoelastic behavior, and detachment dynamics.

3.5 Quantification of Biofilms by Crystal Violet (CV) Staining

1. To validate QCM results, biofilm biomass was quantified using crystal violet (CV) staining. After incubation (typically 24 h), spent medium was carefully removed, and biofilm-coated surfaces were gently washed with sterile water to remove planktonic cells (Fig. 1).

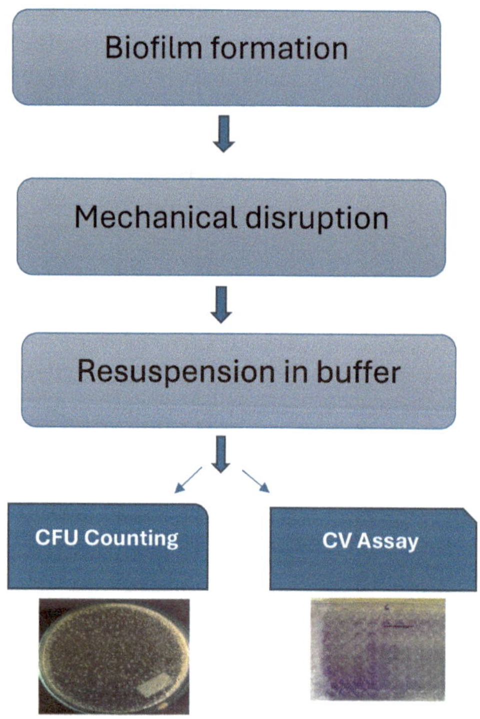

Fig. 1 Quantification of biofilm formation. Schematic representation of the workflow used for biofilm quantification. Following biofilm formation, samples are mechanically disrupted and resuspended in buffer to obtain a homogeneous suspension. Biofilm biomass is then quantified using two complementary methods: colony-forming unit (CFU) counting to determine viable bacterial cells and crystal violet (CV) assay to assess total biofilm biomass. This combined approach enables evaluation of both viable cell density and overall biofilm mass

2. Samples were air-dried, stained with 0.5% CV solution for 20 min at room temperature, and washed thoroughly to remove excess dye (*see* **Note 6**). Bound CV was solubilized with methanol, and absorbance was measured at 570 nm using a microplate reader. Parallel assays in 96-well plates were performed as controls [8].

3.6 Cleaning of QCM Crystals

1. Gold-coated quartz crystals were cleaned using freshly prepared "piranha solution" (3:1 mixture of concentrated H_2SO_4 and 30% H_2O_2) (*see* **Note 7**). Crystals were immersed for ~2 min, rinsed with Milli-Q water, and dried under a nitrogen stream (*see* **Note 8** and Fig. 2).
2. Titanium-coated crystals, due to susceptibility to etching, were cleaned by sequential sonication in acetone and isopropanol (5 min each), followed by rinsing and nitrogen drying (*see* **Note 9**). Each crystal could be reused up to four cycles with this cleaning protocol [9] (*see* **Notes 10** and **11**).

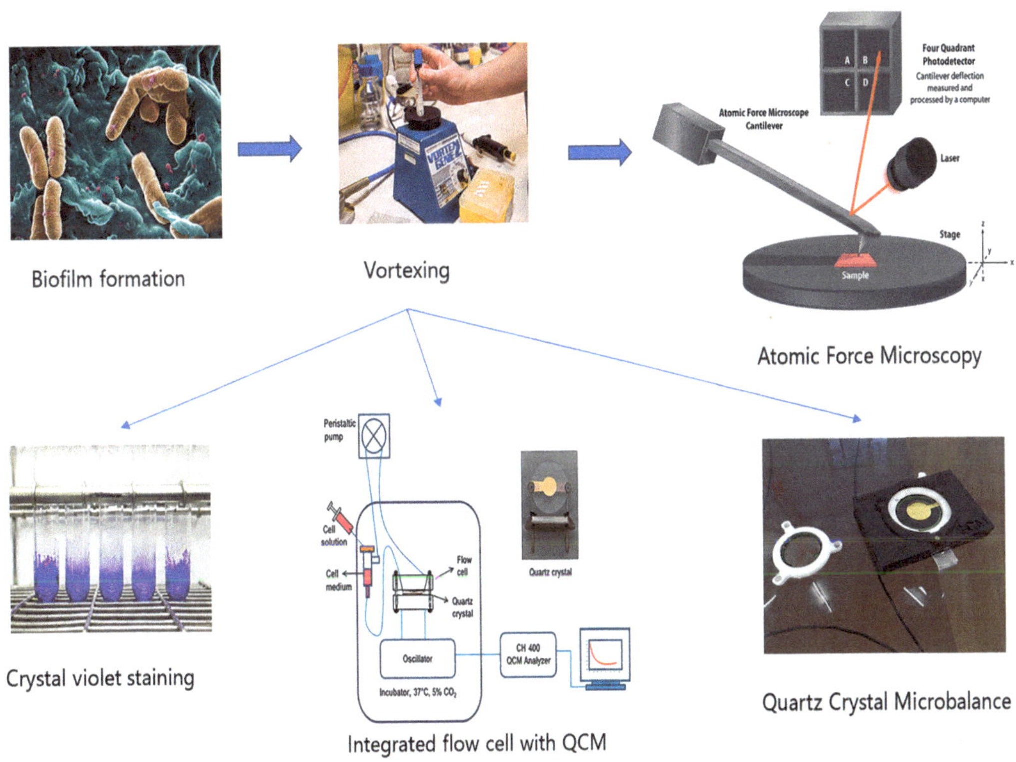

Fig. 2 Schematic representation of the experimental approach used to evaluate biofilm formation and associated biophysical properties. Biofilms are first cultivated under controlled conditions, followed by mechanical disruption through vortexing. Subsequent analyses include structural characterization using atomic force microscopy (AFM), biomass quantification by crystal violet staining, real-time monitoring using an integrated flow cell system coupled with quartz crystal microbalance (QCM), and mass-based biofilm assessment using a quartz crystal microbalance platform. The diagram illustrates the complementary use of microscopic, staining-based, and biophysical techniques for comprehensive biofilm evaluation

4 Notes

1. Use fresh overnight cultures for reproducibility; avoid overgrown or stationary-phase cultures.
2. Ensure complete baseline stabilization before adding inoculum. Bubbles during loading can affect QCM signals.
3. During AFM imaging, maintain a high set-point ratio (Asp/A-free ≈ 0.85) to minimize tip contamination.
4. Scan multiple regions of the biofilm to account for heterogeneity.
5. Flow cell tubing and connectors must be bubble-free to prevent disruption of resonance frequency readings. Maintain aseptic connections at all times.
6. Wash biofilm samples gently before staining to avoid loss of biomass. Always include negative controls (medium only).
7. Piranha solution is highly corrosive and must only be handled in a fume hood with appropriate PPE. Never store prepared solution.
8. Do not treat titanium electrodes with piranha solution; they should be cleaned with organic solvents only.
9. Perform all measurements in replicates (minimum triplicates) to account for biological variability.
10. Software settings (integration time, smoothing, baseline correction) should be standardized across experiments for comparability.
11. Dispose of chemical waste (especially piranha solution and organic solvents) according to institutional safety protocols.

Competing Interest

The author reports no financial conflicts related to the subject matter of this chapter.

References

1. Costerton JW, Lewandowski Z, Caldwell DE, Korber DR, Lappin-Scott HM (1995) Microbial biofilms. Ann Rev Microbiol 49(1):711–745. https://doi.org/10.1146/annurev.mi.49.100195.003431
2. Flemming HC, Wingender J, Szewzyk U et al (2016) Biofilms: an emergent form of bacterial life. Nat Rev Microbiol 14(9):563–575. https://doi.org/10.1038/nrmicro.2016.94
3. Marx KA (2003) Quartz crystal microbalance: a useful tool for studying thin polymer films and complex biomolecular systems at the solution–surface interface. Biomacromolecules 4(5): 1099–1120. https://doi.org/10.1021/bm020116i
4. Reviakine I, Johannsmann D, Richter RP (2011) Hearing what you cannot see and visualizing what you hear: interpreting quartz crystal

microbalance data from solvated interfaces. Anal Chem 83(23):8838–8848. https://doi.org/10.1021/ac201778h

5. Gaboriaud F, Dufrene YF (2007) Atomic force microscopy of microbial cells: application to nanomechanical properties, surface forces and molecular recognition forces. Colloids Surf B: Biointerfaces 54(1):10–19. https://doi.org/10.1016/j.colsurfb.2006.09.014
6. Tolker-Nielsen T, Sternberg C (2011) Growing and analyzing biofilms in flow chambers. Curr Protoc Microbiol, 1B.2.1–1B.2.17
7. Mann EE, Wozniak DJ, Drescher K (2014) Flow-cell biofilm models for single and multi-species studies. Methods Mol Biol 1149:1–16
8. Stepanovic S, Vukovic D, Hola V et al (2000) Quantification of biofilm in microtiter plates: overview of testing conditions and practical recommendations for assessment of biofilm production by staphylococci. APMIS 108(8):659–663. https://doi.org/10.1034/j.1600-0463.2000.d01-131.x
9. Rodahl M, Hook F, Krozer A, Brzezinski P, Kasemo B (1995) Quartz crystal microbalance setup for frequency and Q-factor measurements in gaseous and liquid environments. Rev Sci Instrum 66(7):3924–3930. https://doi.org/10.1063/1.1145396

Chapter 15

Atomic Force Microscopy Approaches for Biofilm Investigation

Prabha Nagaram

Abstract

Biofilms are structured communities of microorganisms that play a crucial role in industrial, environmental, and medical settings. To effectively control and manage them, it is essential to comprehend their development, genetic regulatory mechanisms, interspecies relationships, and reactions to environmental signals. Atomic force microscopy (AFM) provides incredibly high-resolution insights into cellular and subcellular structural and functional features, but its time-consuming nature and small scan size limit the ability to directly connect these detailed observations to the larger macroscale architecture and functional behavior of biofilms. Traditional microscopy techniques have contributed substantially to understanding biofilm architecture and developmental stages; however, most remain largely qualitative and offer limited insight into biofilm substratum interactions. AFM (atomic force microscopy) has emerged as a potent analytical tool capable of providing both qualitative and quantitative information at the nanoscale, enabling detailed characterization of biofilm morphology, mechanical properties, adhesion forces, and surface alterations. The significance of AFM lies in its ability to complement conventional microscopy by offering novel nanoscale insights into biofilm behavior on surfaces, thereby advancing the understanding of microbe–material interactions in environment.

Key words Atomic force microscopy, Nanoscale imaging, Microbial adhesion, Surface interactions

1 Introduction

Biofilms continue to pose major challenges in both medical and industrial settings. The ability of free-floating (planktonic) microorganisms to organize into structured communities in moist environments, along with their heightened tolerance to antimicrobial agents and the growing prevalence of intrinsic resistance, highlights the urgent need for new biofilm control approaches and advanced techniques to better understand their fundamental characteristics [1].

Atomic force microscopy (AFM) has become an important tool in modern microscopy, offering both high-resolution imaging and force measurement capabilities that allow investigation of surface

Abirami K (ed.), *Microbial Biofilm Research: Comprehensive Protocols and Best Practices*, Springer Protocols Handbooks,
https://doi.org/10.1007/978-1-0716-5535-1_15,

properties at the nanoscale [2]. It has been widely applied to generate valuable insights into biofilm architecture, as well as the interaction forces and mechanical features that influence biofilm behavior and their responses to chemical or physical treatment methods [3]. AFM enables examination of entire biofilms as well as their individual components, including bacterial surface structures, extracellular polymeric substances (EPS), and single microbial cells [4].

Atomic force microscopy (AFM) was introduced in 1986 as a member of the scanning probe microscopy family and was rapidly adopted for imaging biological specimens such as DNA, bacteria, viruses, and mammalian cells. AFM is also capable of measuring interaction forces between surfaces, which has obvious implications for the study of microbial adhesion, biofilm cohesion, and surface colonization processes [5]. As a result, AFM has become an important addition to the group of analytical instruments used to study microbial interactions involved in biofilm formation and persistence.

An AFM system is primarily composed of a cantilever, a piezoelectric scanner, a laser diode, a mirror, a photodetector, and a feedback control unit. The laser beam is directed onto the back of the cantilever and reflected toward the photodetector. When the probe tip interacts with the sample surface, the cantilever bends, causing a shift in the position of the reflected laser on the photodetector. This deflection signal is interpreted by the feedback controller to produce high-resolution surface images and quantitative force data displayed on a computer [6].

2 Materials

1. AFM Instrumentation.
   - Atomic force microscope with controller and imaging software.
   - Vibration isolation system.
2. AFM Probes.
   - Soft silicon or silicon nitride cantilevers (contact or tapping mode).
3. Microbial Culture Materials.
   - Biofilm-forming microorganisms.
   - Growth media (LB broth/TSB/nutrient broth).
4. Substrates for Biofilm Growth.
   - Mica sheets, glass coverslips, or silicon wafers.

5. Sample Preparation Materials.
   - Phosphate-buffered saline (PBS) or sterile water.
   - Fixatives (optional: glutaraldehyde or paraformaldehyde).
6. Imaging Accessories.
   - Liquid cell for live biofilm imaging (or desiccator for air imaging).
7. General Laboratory Supplies.
   - Sterile Petri dishes, pipettes, forceps, gloves.

3 Methods

3.1 Sample Preparation

1. Select an appropriate substrate (e.g., glass coverslip, mica, stainless steel, silicon wafer).
2. Clean the substrate thoroughly using ethanol or acetone, followed by rinsing with sterile distilled water.
3. Sterilize the substrate by autoclaving or UV exposure.
4. Inoculate the substrate with the microbial culture and incubate under suitable conditions to allow biofilm formation.
5. To remove loosely attached planktonic cells, carefully wash the substrate with sterile PBS (phosphate-buffered saline) following incubation.
6. For live biofilm imaging, keep the sample hydrated in buffer.
7. For fixed biofilm imaging, fix the biofilm using 2–2.5% glutaraldehyde, followed by gentle rinsing and air-drying (if imaging in air) (*see* **Note 1**).

3.2 Selection and Preparation of AFM Probe

1. Choose an appropriate cantilever based on imaging mode:
   - Soft cantilevers for live, soft biofilms.
   - Stiffer cantilevers for mechanical measurements.
2. For biological samples, the typical spring constant is 0.01–0.6 N/m.
3. Calibrate the cantilever for sensitivity and spring constant prior to imaging (*see* **Note 2**).

3.3 Instrument Setup

1. Switch on the AFM system, laser, and controller.
2. Align the laser beam onto the back of the cantilever.
3. Adjust the photodetector to maximize the signal and center the laser spot.
4. Mount the prepared biofilm sample onto the AFM sample stage (*see* **Note 3**).

5. Select the imaging environment:
   - Air mode.
   - Liquid mode (preferred for hydrated biofilms).

3.4 Approach of the Cantilever to the Sample

1. Use the automated or manual approach function to bring the cantilever tip close to the biofilm surface.
2. Monitor deflection signals carefully to avoid damaging the biofilm.
3. Set appropriate force limits to minimize deformation of soft biofilm structures.

3.5 Imaging Mode Selection

1. Choose the appropriate AFM imaging mode:
   - *Contact mode*—for rigid, fixed biofilms.
   - *Tapping (intermittent contact) mode*—preferred for soft, hydrated biofilms.
   - *Force-volume or peak force mode*—for mechanical property mapping.
2. Optimize scanning parameters such as scan size, scan rate, and set point force (*see* **Notes 4** and **5**) (Fig. 1).

3.6 Image Acquisition

1. Scan the surface at low resolution to locate biofilm regions.
2. Gradually increase resolution for detailed imaging.
3. Record height, phase, and amplitude images.
4. Capture multiple scans at different locations to ensure representative biofilm analysis (*see* **Notes 6** and 7).

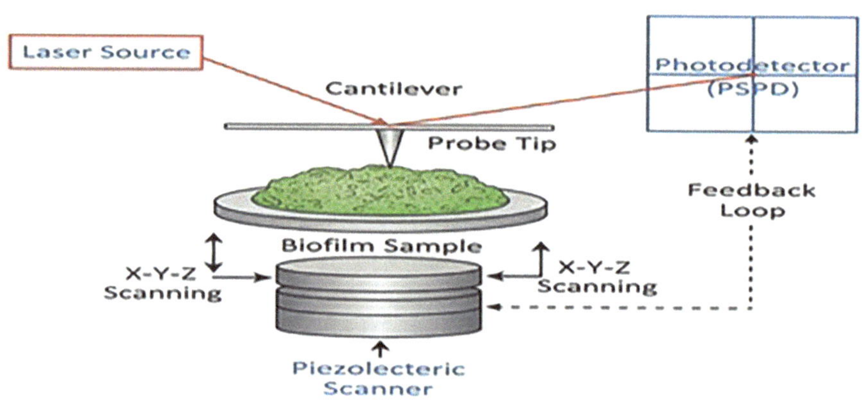

Fig. 1 Schematic diagram of atomic force microscopy

3.7 Force Spectroscopy (Optional)

1. Perform force–distance measurements to evaluate:
   - Adhesion forces.
   - Biofilm stiffness (Young's modulus).
   - Cohesive interactions within the biofilm.
2. Collect force curves across the biofilm surface for statistical reliability (*see* **Notes 8** and **9**).

3.8 Data Processing and Analysis

1. Use AFM software to flatten and correct image artifacts (*see* **Note 10**).
2. Measure surface roughness, biofilm thickness, and topographical features.
3. Analyze force curves to quantify mechanical and adhesive properties.
4. Compare treated vs. untreated biofilms (e.g., after antimicrobial exposure).

3.9 Interpretation of Results

1. Identify biofilm presence through characteristic 3D structures and EPS matrix (*see* **Note 11**).
2. Correlate structural and mechanical properties with biofilm maturity and resistance.
3. Evaluate the effectiveness of antimicrobial or surface treatments.

3.10 Documentation and Reporting

1. Save raw and processed AFM images.
2. Document all experimental parameters (cantilever type, mode, and forces).
3. Report results with scale bars, force values, and statistical analysis.

4 Notes

1. Poor or absent image formation during AFM analysis of biofilms can result from several technical factors. Misalignment of the laser on the cantilever or improper positioning of the photodetector often disrupts signal detection. In some cases, the AFM tip may not be properly engaged with the biofilm surface, or the imaging parameters such as scan rate and feedback settings may be unsuitable. To address these issues, the laser should be carefully realigned to the back of the cantilever and the photodetector signal centered correctly. Proper engagement between the cantilever and the sample must be verified before scanning. Imaging conditions should be optimized by adjusting scan rate, setpoint force, and feedback gains. It is generally advisable to begin with a larger scan area and then gradually reduce it to refine image quality.

2. Damage to biofilms during scanning is another common challenge, particularly because biofilms are soft and highly hydrated structures. Excessive force applied by the tip, the use of contact mode, or high scan speeds can disrupt the matrix and distort structural features. This problem can be minimized by lowering the applied force and reducing the imaging load. For delicate biofilms, tapping mode or peak force mode is preferable over contact mode. Slowing down the scan speed and selecting cantilevers with low spring constants further help preserve structural integrity.
3. Tip contamination or fouling frequently occurs when extracellular polymeric substances (EPS) adhere to the AFM probe, especially in dense or viscous biofilms. A contaminated tip can significantly compromise image quality and force measurements. In such cases, the tip should be cleaned or replaced. Depending on the experimental objective, hydrophobic or chemically functionalized tips may reduce unwanted adhesion. Limiting prolonged contact between the tip and the biofilm surface and gently rinsing the sample to remove excess EPS can also improve performance.
4. Excessive background noise or imaging artifacts may arise from environmental vibrations, electrical disturbances, or improper feedback control. Ensuring proper vibration isolation and acoustic shielding is critical for high-resolution imaging. The AFM system should be properly grounded to prevent electrical interference. Fine-tuning feedback gain and scan parameters, along with maintaining a stable temperature environment, will further enhance signal stability.
5. Poor resolution may be attributed to a worn or blunt tip, incorrect cantilever selection, or uneven biofilm thickness. Replacing the cantilever with a sharp probe designed for biological applications often improves image clarity. Selecting appropriate probes for soft samples and focusing on thinner, more uniform regions of the biofilm can also enhance resolution. Reducing scan size typically increases detail and spatial resolution.
6. Unstable force–distance curves are usually linked to inaccurate cantilever calibration, sample drift, hydration instability, or tip slippage on soft surfaces. Recalibrating cantilever sensitivity and spring constant is essential for reliable force measurements. Maintaining a consistent liquid environment and minimizing evaporation help reduce drift. Using slower approach and retraction speeds and increasing data averaging can stabilize force curves.

7. Sample drift during imaging commonly results from temperature fluctuations or inadequate attachment of the biofilm to the substrate. Allowing the instrument to thermally equilibrate before imaging and firmly securing the sample holder can reduce movement. During liquid-mode experiments, evaporation should be minimized to maintain stability.
8. Engagement difficulties may occur if the Z-piezo range is incorrectly set or if the sample surface is uneven. Adjusting the Z-range appropriately and initiating the approach on a relatively flat area of the biofilm can improve engagement. When available, optical microscope guidance can assist in locating suitable regions for scanning.
9. Low reproducibility of results is often associated with the inherent heterogeneity of biofilms and variability in cantilever properties. To improve reliability, multiple scans should be performed at different locations on the biofilm. Using identical probe types for comparative experiments and standardizing biofilm growth and preparation conditions are essential for consistency.
10. Data misinterpretation is another concern, particularly when distinguishing true topographical features from imaging artifacts. Over-processing or excessive flattening of images can distort structural information. To avoid this, height, phase, and amplitude images should be evaluated collectively. Minimal image processing is recommended, and AFM findings should be validated using complementary techniques such as scanning electron microscopy (SEM) or confocal laser scanning microscopy (CLSM).
11. Several precautions are necessary when working with biofilms using AFM. Unintentional drying of live biofilms should be avoided, as it alters structural and mechanical properties. Imaging parameters must always be documented for reproducibility. Cantilevers should be handled carefully to prevent damage. Finally, control experiments using biofilm-free substrates are important to distinguish true biological features from surface artefacts.

Competing Interest

The author(s) confirm that there are no financial competing interests associated with this chapter.

References

1. Chatterjee S, Biswas N, Datta A, Dey R, Maiti P (2014) Atomic force microscopy in biofilm study. Microscopy 63(4):269–276. Oxford University Press. This article reviews AFM's advantages over SEM for biofilm surface characterization under ambient conditions. (OUP Academic)
2. Hobley L, Henriques M (2010) Application of AFM from microbial cell to biofilm. J Microsc. Wiley Periodicals, Inc. This review outlines how AFM has been used to study microbial cells and biofilms at multiple scales
3. Steele A, Goddard DT, Beech IB (1994) An atomic force microscopy study of the biodeterioration of stainless steel in the presence of bacterial biofilms. Int Biodeterior Biodegradation 34(1):35–46. https://doi.org/10.1016/0964-8305(94)90018-3
4. Holscher P, Ducker WA, Sagstetter M (2017) Atomic force microscopy studies of bioprocess engineering surfaces—imaging, interactions and mechanical properties mediating bacterial adhesion. Biofouling 33(6):523–544. A review highlighting AFM for imaging and quantifying bacterial adhesion and mechanics at interfaces
5. Ahimou F, Semmens MJ, Novak PJ, Haugstad G (2007) Biofilm cohesiveness measurement using a novel atomic force microscopy methodology. Appl Environ Microbiol 73(9):2897–2904
6. Binnig G, Quate CF, Gerber C (1986) Atomic force microscope. Phys Rev Lett 56:930–933. The original paper introducing AFM as a scanning probe technique

Chapter 16

Flow Cell Systems: Allow Real-Time Observation of Biofilm Formation Under Controlled Flow Conditions

Rajesh Palanisamy

Abstract

Flow cell systems represent the gold standard for real-time observation of biofilm formation under physiologically relevant hydrodynamic conditions. This chapter provides a comprehensive guide to designing, assembling, operating, and analyzing biofilms in flow cell systems, enabling researchers to study biofilm development dynamics, architecture, and response to environmental perturbations with unprecedented temporal and spatial resolution. We detail protocols for flow cell fabrication, sterilization, inoculation with common biofilm-forming bacteria (*Pseudomonas aeruginosa*, *Staphylococcus aureus*), establishment of laminar flow profiles, and non-destructive imaging using confocal laser scanning microscopy (CLSM) and time-lapse video microscopy. Special attention is given to quantitative image analysis workflows for measuring biomass accumulation, microcolony development, extracellular polymeric substance (EPS) production, and biofilm detachment events. Real-world examples demonstrate how flow cells reveal stage-specific biofilm maturation, quorum sensing activation, antimicrobial tolerance development, and shear stress-induced architectural remodeling. This chapter equips microbiologists, bioengineers, and biofilm researchers with practical knowledge and troubleshooting strategies for producing reproducible, publication-quality biofilm datasets under controlled flow conditions.

Key words Flow cell systems, Biofilm formation, Real-time imaging, Hydrodynamic conditions, Confocal microscopy, Quantitative image analysis, *Pseudomonas aeruginosa*, *Staphylococcus aureus*, Laminar flow, Biofilm architecture

1 Introduction

Biofilms represent the predominant microbial lifestyle, with over 80% of bacterial infections involving biofilm communities protected by extracellular polymeric substances (EPS) [1, 2]. Traditional static biofilm models fail to replicate the hydrodynamic shear forces, nutrient gradients, and oxygen stratification characteristic of natural and industrial biofilm environments [5]. Flow cell systems overcome these limitations by enabling real-time observation of biofilm formation under precisely controlled flow conditions,

Abirami K (ed.), *Microbial Biofilm Research: Comprehensive Protocols and Best Practices*, Springer Protocols Handbooks,
https://doi.org/10.1007/978-1-0716-5535-1_16,

Table 1
Representative RT-qPCR primers for biofilm development and virulence genes

Gene	Function	Forward primer (5′-03′)	Reverse primer (5′-03′)	Tm (°C)	Amplicon (bp)	Expected FC
pelA	EPS/Psl biosynthesis	GCTACGGCTT CGACAACGAT	CGATGCCAGTA GCCGATCTT	60	156	↑↑
lasR	Quorum sensing (Las system)	ACG TTCAGCGACG ACATCG	TGACGATCACG TCGATCTCA	58	142	↑↑
rhlR	Quorum sensing (Rhl system)	GATCGACA TCGACGTGC TGA	GCGGTGATCA TCTGCGTC TG	60	187	↑↑
algD	Alginate biosynthesis	CCATCGTCA TCGACATCGAC	AGGTCGGTG TGAACGGA TTTG	59	134	↑
oprI	Outer membrane protein	ATGACCA TGACCACGC TGAA	TGTAGACCATG TAGTTGAGG T	60	123	↑
16S rRNA	Bacterial housekeeping control	AGAGTTTGATCC TGGCTCAG	ACGGCTACC TTGTTACG ACT	58	200	=

providing unprecedented insight into biofilm development dynamics [3, 6] (Table 1).

Flow cells maintain laminar flow regimes (Reynolds numbers <2000) that mimic physiological shear stresses encountered in the urinary tract, oral cavity, and industrial pipelines [4]. Transparent parallel plate or channel designs allow continuous non-destructive imaging via light microscopy, epifluorescence, or confocal laser scanning microscopy (CLSM) [7]. Researchers can observe initial bacterial adhesion, microcolony formation, EPS matrix development, mushroom-shaped architecture emergence, and dispersal events occurring over minutes to days [8].

1.1 Key Advantages of Flow Cell Systems Include:

- *Controlled hydrodynamics:* Precise regulation of flow rate (0.01–10 mL/min), shear stress (0.01–10 dyn/cm^2), and residence time enables mimicry of diverse biofilm microenvironments [9].
- *Gradient formation:* Establishment of nutrients, oxygen, and chemical concentration gradients across biofilm depth reflect physiological nutrient limitation and heterogeneity [10].

- *Perturbation capability:* Real-time introduction of antibiotics, quorum sensing inhibitors, or environmental stressors enable dynamic response studies [11].
- *Quantitative analysis:* Automated image processing for biomass, thickness, and structural parameter quantification enables statistical comparison of biofilm phenotypes [9].

This chapter synthesizes current best practices for flow cell biofilm research, guiding researchers from system assembly through data analysis. Special focus addresses common challenges including bubble formation prevention, uniform inoculation, and reproducible flow profiles essential for publication-quality biofilm studies.

2 Materials

2.1 Flow Cell Fabrication and Assembly

- Polydimethylsiloxane (PDMS) elastomer kit (Sylgard 184, Dow Corning).
- Glass microscope coverslips (No. 1.5, 24 × 60 mm).
- Acrylic sheets (3 mm thick) for flow cell housing.
- Silicone tubing (1/16″ ID, Tygon).
- Luer-lock fittings and barbed connectors.
- Peristaltic pump (e.g., Watson-Marlow 120S).
- Bubble trap (custom or commercial, e.g., Cole-Parmer).
- Stainless steel needles (21–25 gauge) for media injection.

2.2 Bacterial Strains and Culture Media

- *Pseudomonas aeruginosa* PAO1 or PA14 (ATCC).
- *Staphylococcus aureus* Newman or USA300 (clinical isolates).
- Flow cell medium: 1× M9 minimal salts +0.4% glucose +0.1% Casamino acids.
- LB broth for preculture.
- Phosphate-buffered saline (PBS), pH 7.4.

2.3 Imaging and Analysis Equipment

- Inverted confocal laser scanning microscope (e.g., Zeiss LSM 880).
- 40×/1.2 NA water immersion objective.
- Environmental chamber (37 °C, 5% CO_2).
- ImageJ/Fiji with BiofilmQ plugin [12].
- Imaris or Volocity for 3D reconstruction. Imaris or Volocity are software packages used for 3D visualization and analysis of microscopy images.
- High-resolution CMOS camera for time-lapse imaging.

2.4 Staining and Probes

- SYTO 9/propidium iodide LIVE/DEAD BacLight kit (Thermo Fisher).
- Wheat germ agglutinin-Alexa Fluor 488 conjugate (EPS staining).
- FITC-conjugated lectins (mannose/galactose-specific).
- Calcofluor white (β-1,3-glucans).

2.5 Flow Control and Perfusion

- Syringe pumps (0.01–10 mL/min precision).
- Media reservoirs (glass bottles, 500 mL).
- Inline filters (0.22 μm).
- Flow meters (optional, for precise calibration).

3 Methods

3.1 Flow Cell Design and Fabrication

1. *PDMS channel fabrication:* Design parallel plate channel (width 1–4 mm, height 0.1–1 mm, length 20–40 mm) using AutoCAD or equivalent software. Create SU-8 photoresist mold via photolithography [11]. Mix PDMS (10:1 base: curing agent ratio), pour over mold, cure at 60 °C for 4 h, and plasma bond to glass coverslip to create irreversible seal (*see* Fig. 1) [13]. Verify channel geometry using optical profilometry (*see* **Note 1**).
2. *Housing assembly:* Mount PDMS channel between acrylic plates with inlet/outlet ports drilled precisely using calibrated drill guides. Use silicone gaskets to seal without distorting channel dimensions. Verify channel height using interference reflection microscopy to confirm uniform geometry [14].
3. *Sterilization:* Autoclave acrylic housing (121 °C, 20 min, 15 psi). Flow 70% ethanol through PDMS channel (5 mL/min, 30 min) to disinfect interior surfaces, followed by sterile water rinse (5 mL/min, 30 min) and air drying under laminar flow hood [15].

3.2 Bacterial Preculture and Inoculation

1. *Overnight culture:* Inoculate single *P. aeruginosa* PAO1 colony (stored on glycerol stocks or agar plates) into 5 mL LB medium. Grow at 37 °C with shaking at 200 rpm for 16 h to reach stationary phase ($OD_{600} = 0.8–1.0$) (*see* Figs. 1 and 2).
2. *Biofilm inoculum preparation:* Wash cells twice in sterile flow cell medium (3000× *g*, 5 min, 4 °C) to remove residual LB broth. Resuspend cell pellet to $OD_{600} = 0.05$ in flow cell medium (approximately 5×10^7 CFU/mL) [16].
3. *Static attachment phase:* Inject 200 μL inoculum into flow cell using sterile syringe. Invert flow cell (coverslip facing down, PDMS channel facing up) and incubate statically at 37 °C for 1 h to allow initial bacterial adhesion and microcolony formation (*see* **Note 2**) [13].

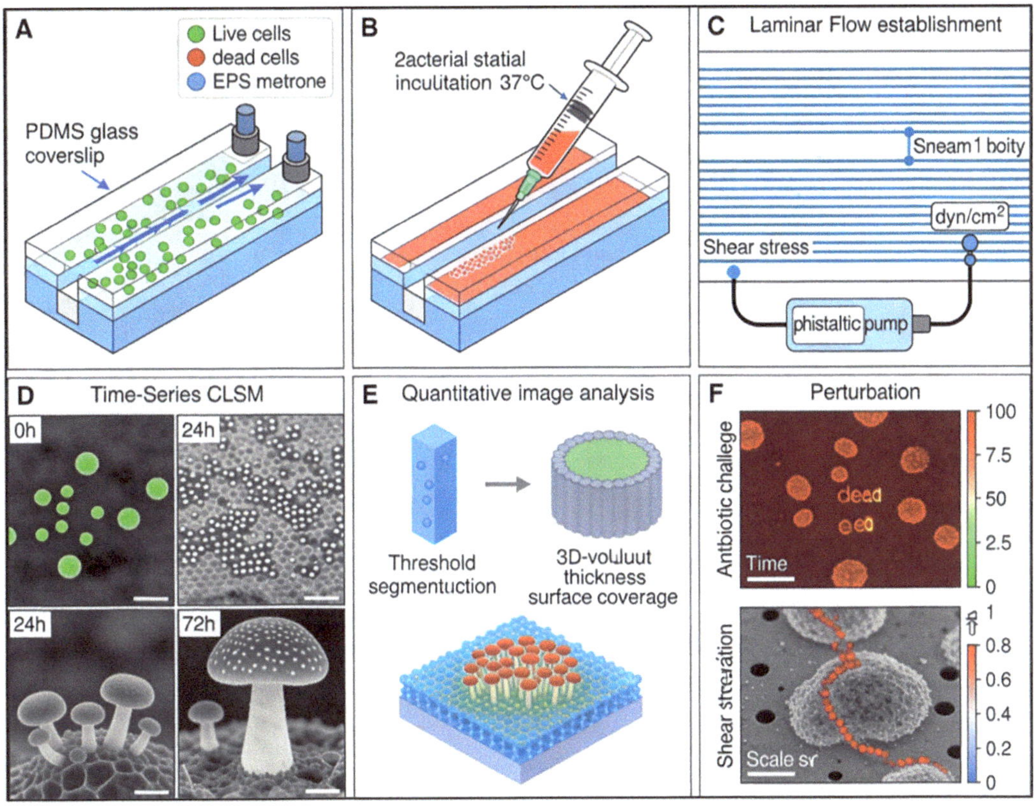

Fig. 1 Comprehensive flow cell biofilm experimental workflow. (**a**) Flow cell assembly and sterilization procedures. (**b**) Bacterial inoculation and static attachment phase (0–1 h). (**c**) Laminar flow establishment with real-time confocal imaging setup. (**d**) CLSM time-series showing biofilm development stages (0, 24, 48, 72 h post-inoculation). (**e**) Quantitative image analysis pipeline for extraction of biofilm structural parameters. (**f**) Perturbation experiments demonstrate antibiotic challenge and shear stress response effects on biofilm architecture

3.3 Flow Cell Operation and Flow Profile Establishment

1. *Connect perfusion system:* Attach inlet to sterile media reservoir (prewarmed to 37 °C) via silicone tubing, bubble trap, and peristaltic pump. Connect outlet to sterile waste collection bottle. Prime entire system with sterile flow cell medium at 0.01 mL/min for 10 min to remove air bubbles and establish equilibrium (*see* **Note 3**) (*see* Fig. 1) [17].
2. *Initiate laminar flow:* After static attachment phase, carefully orient flow cell to horizontal position with inlet upstream. Start peristaltic pump at 0.1 mL/min (generating shear stress ~0.02 dyn/cm^2). Verify uniform laminar flow profile using fluorescein dye streaks (inject dye solution at inlet, observe straight streamlines indicating uniform velocity gradient) or particle image velocimetry (PIV) if available [14].

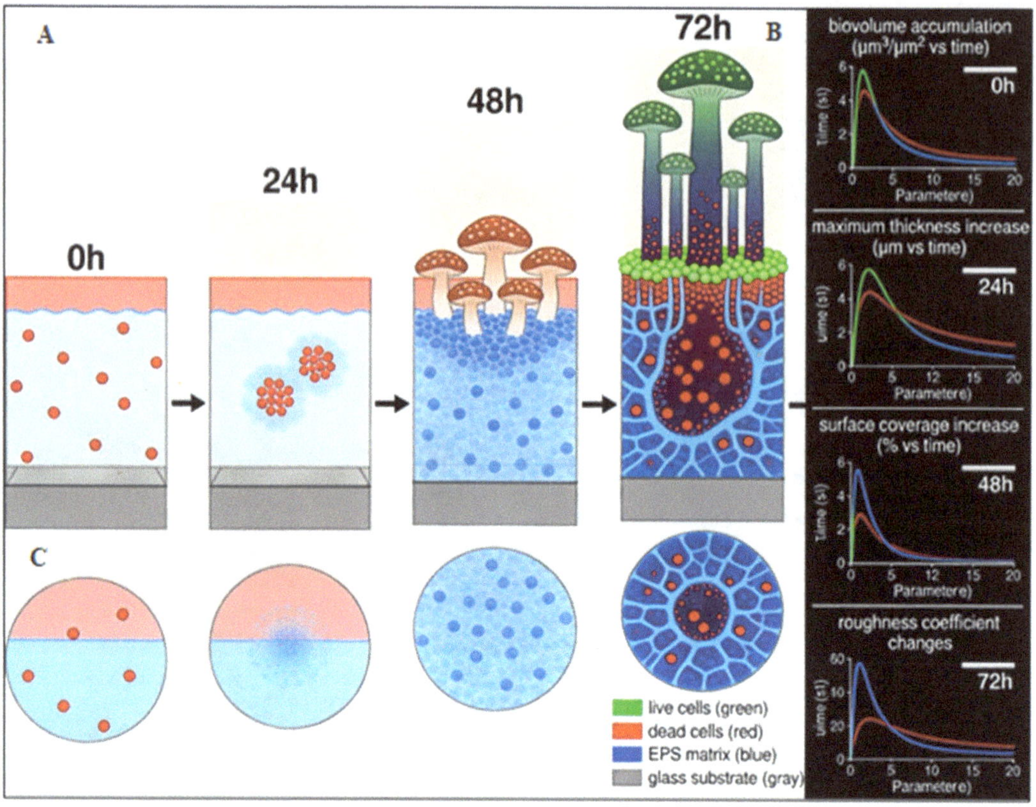

Fig. 2 Representative *Pseudomonas aeruginosa* PAO1 biofilm development in flow cell system. (**a**) Time course images showing biofilm formation at 0, 24, 48, and 72 h post-inoculation. (**b**) Quantitative parameters derived from COMSTAT2 image analysis: biovolume accumulation ($\mu m^3/\mu m^2$), maximum thickness evolution (μm), and surface coverage progression (%). (**c**) Three-dimensional reconstruction of mature biofilm at 72 h showing characteristic mushroom-shaped architecture with EPS matrix (green, WGA staining). LIVE/DEAD viability staining reveals heterogeneous cell population with viable cells (green, SYTO 9) predominantly at biofilm periphery and dead/dying cells (red, propidium iodide) concentrated in interior oxygen-limited microenvironments

3. *Biofilm development monitoring:* Begin confocal microscopy imaging immediately after flow initiation (0 h timepoint). Acquire z-stacks (1 μm optical sections) every 15–60 min for 24–72 h depending on biofilm growth kinetics. Maintain constant temperature (37 °C ± 0.5 °C) and flow rate throughout experiment (*see* Fig. 2) [18].

3.4 Real-Time Imaging Protocols

3.4.1 Confocal Laser Scanning Microscopy (CLSM)

1. *Live/dead staining:* Perfuse LIVE/DEAD BacLight stain (3 μM SYTO 9 nucleic acid stain +30 μM propidium iodide) diluted in flow cell medium through flow cell for 15 min at 0.1 mL/min. Stop flow, let stain incubate for 5 min in dark. Resume flow cell medium rinse (5 min at 0.1 mL/min) to remove unincorporated dye (*see* Fig. 2) [3].

2. *EPS matrix visualization:* Perfuse wheat germ agglutinin-Alexa Fluor 488 (5 μg/mL; binds N-acetyl-D-glucosamine in polysaccharides) + calcofluor white (0.01%; binds β-1,3-glucans in Polysaccharide synthesis locus) for 30 min at 0.05 mL/min. Rinse with flow cell medium (15 min at 0.1 mL/min) [19].
3. *Imaging parameters:* Acquire images using 488 nm (SYTO 9, WGA-Alexa488) and 561 nm (propidium iodide) laser lines. Set z-step to 1 μm for optical sectioning (approximately 50–60 μm total depth). Scan resolution 1024 × 1024 pixels, frame averaging 2× and using a 40× water-immersion objective (1.2 NA, working distance 0.23 mm). Acquire 5–10 positions per channel to ensure spatial representation (*see* **Note 4**) [20].

3.4.2 Time-Lapse Phase-Contrast Microscopy

1. *Unstained imaging:* Acquire phase-contrast images every 5 min continuously for 48 h to track microcolony coalescence dynamics without phototoxicity. Phase contrast enables visualization of cell density changes and architecture remodeling without requiring fluorescent staining [21].
2. *Automated stage control:* Program motorized x-y stage to revisit identical fields-of-view (FOVs) across experiments with ±1 μm lateral positioning accuracy and ± 2 μm axial accuracy. Store stage positions in experiment metadata for downstream image registration and quantification [22].

3.5 Quantitative Biofilm Analysis

3.5.1 Image Processing and Biovolume Quantification

1. *COMSTAT2 analysis* [23]: Export CLSM z-stack series (SYTO 9 channel) as 8-bit grayscale images. Import into COMSTAT2 software. Threshold images to segment bacterial cells (typically threshold = 50–100 arbitrary units). Calculate biofilm structural parameters (e.g., biomass, average thickness, roughness coefficient, and surface-to-volume ratio) for each sample (*see* **Note 5**) (Table 2):
   - *Total biovolume:* Sum of voxels occupied by biofilm divided by substratum area ($\mu m^3/\mu m^2$).
   - *Average thickness:* Mean z-position of biofilm voxels (μm).
   - *Maximum thickness:* Highest z-position containing biofilm (μm).
   - *Surface coverage:* Percentage of substratum occupied by biofilm (%) [23].
   - *Roughness coefficient:* Quantifies surface irregularity; values >1 indicate complex architecture [23].
2. *Biofilm workflow* [12]: Use advanced image analysis plugin for automated feature extraction. Segment SYTO 9 and propidium iodide channels separately to distinguish viable vs. dead cells (*see* Fig. 2). Calculate (*see* **Note 6**):

- *Cell density:* Viable cell counts per biofilm volume (cells/μ m^3).
- *EPS coverage:* Percentage of biofilm volume occupied by extracellular polymeric substances (%).
- *Microcolony size distribution:* Histogram of individual microcolony volumes (μm^3) [12].
- *Spatial heterogeneity:* Coefficient of variation in cell density across biofilm depth.

3.5.2 Structural Parameter Calculation

1. *Substratum coverage:* Calculate percentage of substratum (glass coverslip surface) occupied by biofilm using threshold-based binary segmentation. Divide occupied pixels by total substratum pixels, multiply by 100 (Table 2) (*See* Fig. 2) [24].
2. *Roughness coefficient:* $Sa = (1/n)\, \Sigma|Zi - Zmean|/Zmean$ where Zi = z-position of each biofilm voxel, $Zmean$ = average biofilm thickness, n = number of voxels. Values approaching 0 indicate

Table 2
Flow cell operating parameters and expected biofilm characteristics

Parameter	Range	Units	Typical value	Notes
Channel width	1–4	mm	2	Wider channels reduce shear stress; optimize for biofilm morphology
Channel height	0.1–1	mm	0.5	Thinner channels increase shear stress; balance optical clarity
Flow rate	0.01–1	mL/min	0.1	Adjust for desired Reynolds number and shear conditions
Shear stress	0.01–05	dyn/cm^2	0.1–0.5	Physiological range varies by natural biofilm site
Reynolds number	<2000	Dimensionless	100–500	Maintains laminar flow regime critical for interpretation
Residence time	1–60	min	10–30	Controls nutrient limitation gradient and mass transfer
Initial attachment time	0.5–2	h	1	Static incubation before flow initiation
Imaging interval	5–60	min	15	Shorter intervals capture rapid coalescence events
Total culture duration	24–168	h	48–072	Mature biofilm architecture develops by 48 h
Temperature	35–38	°C	37	Maintain ±0.5 °C stability throughout experiment

smooth, flat biofilms; values >1 indicate highly irregular, mushroom-like architecture [23].

3. *Maximum thickness:* Determine highest z-position containing biofilm signal (substratum = z_0). Subtract z_0 from maximum biofilm voxel z-position to calculate thickness in μm (*see* Fig. 2) (Table. 2) [23].

3.6 Perturbation Experiments

1. *Antibiotic challenge:* After 24 h of biofilm development under standard flow conditions, switch media supply to flow cell medium containing ciprofloxacin (1–10 μg/mL final concentration) or tobramycin (50–200 μg/mL final concentration). Continue flowing perturbed medium at 0.1 mL/min for 24 h. Acquire images every 30 min to capture biofilm structural changes and cell death kinetics (*see* Fig. 2) [25].
2. *Quorum sensing inhibition:* Perfuse C4-HSL competitive analogs (synthetic furanones, 0–100 μM) to disrupt *lasR/rhlR*-dependent quorum sensing signaling [26]. Monitor real-time effects on EPS production (WGA staining), cell aggregation (CLSM), and gene expression (combine with parallel RT-qPCR on biofilm samples collected at timepoints) (Table 1) [27].
3. *Shear stress variation:* Increase flow rate five- to tenfold (from 0.1 mL/min to 0.5–1.0 mL/min) to examine biofilm detachment dynamics [28]. Acquire high-frequency time-lapse imaging (every 1–5 min) during flow increase to capture cell sloughing events. Calculate detachment rate as viable cells lost per minute (*see* **Note 7**) (Table 2) (*see* Figs. 1 and 2) [18].

4 Notes

1. *Bubble prevention* (critical for flow reproducibility): Bubbles disrupt laminar flow profiles and create stagnant zones. Use bubble traps at inlet and outlet. Degas all media using vacuum (15 min, <100 mbar) before use to prevent dissolved gas nucleation. Prime tubing and flow cell slowly (0.01 mL/min for initial 20 min) to gradually displace air. If bubbles form during experiment, stop flow immediately, gently tap tubing, resume flow at reduced rate (0.05 mL/min) until bubble clears (Table 2) [17, 29].
2. *Channel clogging:* Monitor pressure drops across flow cell (install pressure gauge at inlet and outlet). Elevated pressure indicates accumulating biomass or debris. To prevent clogging, filter all media through 0.22 μm syringe filters before use, maintain low initial inoculum density (OD_{600} < 0.05 prevents rapid channel blockage), and avoid adding viscosity-increasing

agents (glycerol, viscous polymers) that clog channels (Table 2).

3. *Uniform inoculation* (critical for reproducibility): Static incubation for 1 h ensures even initial adhesion across channel width. Alternatively, inject inoculum at 0.01 mL/min flow for 30 min (minimizes mechanical disturbance while distributing cells). For very wide channels (width > 4 mm), inject from multiple inlet ports simultaneously (Table 2) [16].
4. *Photobleaching minimization* (preserves image quality): Limit laser power to <2% of maximum to minimize phototoxicity and dye photobleaching. Use sequential scanning (alternating 488 nm and 561 nm acquisition) rather than simultaneous multi-channel imaging. Rotate imaging positions (acquire FOV1, then FOV2, return to FOV1 after 20 min) to allow fluorophore recovery [30].
5. *Flow calibration* (ensures experimental reproducibility): Verify laminar flow (Re < 2000) using dye streaks or micro-PIV. Calculate shear stress using formula: $\tau = 6\mu Q/(wh^2)$, where μ = fluid viscosity (0.001 Pa·s at 37 °C), Q = volumetric flow rate (mL/min), w = channel width (mm), and h = channel height (mm). For 2 mm wide × 0.5 mm tall channel at 0.1 mL/min, $\tau = 6 \times 0.001 \times (0.1/60) / (2 \times 0.5^2) \approx 0.02$ dyn/cm^2 (Table 2) [31].
6. *Quality control metrics* (before accepting experimental data):
   - Uniform initial attachment: >80% surface coverage within 2 h post-inoculation.
   - Stable flow profile: velocity variation <5% measured across channel vertical profile.
   - Reproducible biovolume growth: coefficient of variation <20% between replicate channels.
   - CLSM z-stack image quality: signal-to-noise ratio > 10:1.
   - Image registration accuracy: ±1 μm lateral and ± 2 μm axial alignment across 24–72 h timepoints.
7. *Common pitfalls to avoid:*
   - Overly aggressive flow during initial 12 h detaches microcolonies before biofilm establishes; begin with 0.01 mL/min, and gradually increase to 0.1 mL/min by 6 h [13].
   - Insufficient imaging frequency misses rapid microcolony coalescence events (coalescence occurs within 10–30 min); acquire images every 5–15 min for early timepoints [18].
   - Temperature fluctuations (> ± 0.5 °C) alter bacterial metabolism, growth rates, and media viscosity; use environmental chamber with PID temperature controller.

- Inadequate bubble removal by priming disrupts laminar flow; invest 30 min in careful bubble-free priming before inoculation.
- Excessive laser power (>5%) causes phototoxicity, reactive oxygen species generation, and altered bacterial behavior.

Competing Interest

The author declares that they have no financial conflicts of interest.

References

1. Hall-Stoodley L, Costerton JW, Stoodley P (2004) Bacterial biofilms: from the natural environment to infectious diseases. Nat Rev Microbiol 2:95–108. https://doi.org/10.1038/nrmicro821
2. Costerton JW, Stewart PS, Greenberg EP (1999) Bacterial biofilms: a common cause of persistent infections. Science 284:1318–1322. https://doi.org/10.1126/science.284.5418.1318
3. Tolker-Nielsen T, Molin S (2000) Spatial organization of microbial biofilm communities. FEMS Microbiol Lett 186:1–8. https://doi.org/10.1111/j.1574-6968.2000.tb09068.x
4. Heydorn A, Ersbøll BK, Hentzer M et al (2000) Experimental reproducibility in flow-cell biofilms. Microbiology 146:2409–2415. https://doi.org/10.1099/00221287-146-10-2409
5. Purevdorj-Goyil B, Ramsey ME, Hozalski RM et al (2002) Dispersal of biofilms by secreted, matrix-degrading, bactericidal Pseudomonas aeruginosa enzymes. Mol Microbiol 45:1207–1216. https://doi.org/10.1046/j.1365-2958.2002.03116.x
6. Sternberg C, Christensen AB, Johansen CR et al (1999) Construction of a reporter gene system for biosensors and its application to whole cell biosensors for biofilm formation. Methods Enzymol 310:203–213. https://doi.org/10.1016/S0076-6879(99)10016-9
7. Lawrence JR, Korber DR, Hoyle BD et al (1991) Optical sectioning of microbial biofilms. J Bacteriol 173:6558–6567. https://doi.org/10.1128/jb.173.20.6558-6567.1991
8. Wimpenny JWT, Colasanti R (1997) A unifying hypothesis for the structure of microbial biofilms based on cellular individual-based modelling. FEMS Microbiol Ecol 22:1–16. https://doi.org/10.1111/j.1574-6941.1997.tb00354.x
9. Møller S, Korber DR, Wolfaardt GM et al (1997) The morphology of biofilms. Appl Environ Microbiol 63:2742–2745. https://doi.org/10.1128/aem.63.7.2742-2745.1997
10. Picioreanu C, van Loosdrecht MCM, Heijnen JJ (1998) Mathematical modeling of biofilm structure with a hybrid differential-discrete cellular automaton approach. Biotechnol Bioeng 58:101–116. https://doi.org/10.1002/(SICI)1097-0290(19980405)58:1<101::AID-BIT11>3.0.CO;2-M
11. Zhang TC, Bishop PL (1994) Experimental determination of the dissolved oxygen diffusivity in biofilms by microelectrode. Water Res 28:2281–2290. https://doi.org/10.1016/0043-1354(94)90042-6
12. Hartmann A, Klusmann S, Seifert J et al (2021) BiofilmQ 2.0: a platform for comprehensive quantitative analysis of biofilm images. Microorganisms 9:1612. https://doi.org/10.3390/microorganisms9081612
13. O'Toole GA, Kolter R (1998) Initiation of biofilm formation by Pseudomonas fluorescens WCS365 proceeds through a phenotypic transition. Mol Microbiol 28:449–461. https://doi.org/10.1046/j.1365-2958.1998.00799.x
14. Stoodley P, Cargo R, Rupp CJ et al (2002) Biofilm formation and dispersal: the increasing role of hydrodynamic shear forces. Ann Rev Microbiol 56:187–209. https://doi.org/10.1146/annurev.micro.56.012302.160705
15. Heydorn A, Nielsen AT, Hentzer M et al (2000) Quantification of biofilm structures by the novel computer program COMSTAT. Microbiology 146:2395–2407. https://doi.org/10.1099/00221287-146-10-2395
16. Nadell CD, Ricaurte D, Yan J et al (2017) Flow environment and matrix structure interact to

determine spatial heterogeneity in biofilms. ISME J 11:1245–1256. https://doi.org/10.1038/ismej.2017.6

17. Rusconi R, Guasto JS, Stocker R (2014) Microfluidic chambers for optical studies of cell populations and biofilms. Nat Protoc 9: 125–135. https://doi.org/10.1038/nprot.2013.155
18. Johns MB, Hanke A, Peters AC et al (2023) Dynamic biofilm development under flow: from single cells to structured communities. Curr Opin Microbiol 72:102264. https://doi.org/10.1016/j.mib.2022.102264
19. Persat A, Nadell CD, Kim MK et al (2014) The mechanical world of bacteria. Cell 158:1436–1450. https://doi.org/10.1016/j.cell.2014.08.005
20. Renault T, Tremblay J, Puttaswamy N et al (2020) Bacterial biofilm shows persistent resistance to liquid wetting. Langmuir 36:2672–2681. https://doi.org/10.1021/acs.langmuir.9b03891
21. Klapper I, Dockery J (2010) Mathematical description of microbial biofilms. SIAM Rev 52:221–265. https://doi.org/10.1137/080739720
22. Flemming H-C, Wingender J (2010) The biofilm matrix. Nat Rev Microbiol 8:623–633. https://doi.org/10.1038/nrmicro2415
23. Bridier A, Briandet R, Thomas V, Dubois-Brissonnet F (2011) Resistance of bacterial biofilms to disinfectants: a review. Biofouling 27:1017–1032. https://doi.org/10.1080/08927014.2011.626899
24. Harmsen M, Lappann M, Knochel S, Molin S (2010) Role of Pseudomonas aeruginosa surface-associated hypoxia in antibiotic resistance. Antimicrob Agents Chemother 54: 4765–4771. https://doi.org/10.1128/AAC.00491-10
25. Hogardt M, Trebesius K, Geiger AM et al (2000) Genome size variation among Pseudomonas aeruginosa strains and its correlation with virulence. J Bacteriol 182:4650–4656. https://doi.org/10.1128/JB.182.16.4650-4656.2000
26. Hentzer M, Wu H, Andersen JB et al (2003) Attenuation of Pseudomonas aeruginosa virulence by quorum sensing inhibitors. EMBO J 22:3803–3815. https://doi.org/10.1093/emboj/cdg366
27. Rasmussen TB, Skindersoe ME, Bjarnsholt T et al (2005) Identity and effects of quorum-sensing inhibitors produced by Penicillium species. Microbiology 151:1325–1340. https://doi.org/10.1099/mic.0.27715-0
28. Stoodley P, Sauer K, Davies DG, Costerton JW (2002) Biofilms as complex differentiated communities. Ann Rev Microbiol 56:187–209. https://doi.org/10.1146/annurev.micro.56.012302.160705
29. Lebeaux D, Chauhan A, Rendueles O, Beloin C (2013) From in vitro to in vivo models of bacterial biofilm-related infections. Pathogens 2:288–356. https://doi.org/10.3390/pathogens2020288
30. Scher K, Römling U, Yaron S (2005) Effect of heat, acidification, and chlorination on Salmonella enterica serovar Typhimurium cells in a biofilm formed at the air–liquid interface. Appl Environ Microbiol 71:1727–1732. https://doi.org/10.1128/AEM.71.4.1727-1732.2005
31. Wagner VE, Bushnell D, Passador L et al (2003) Microarray analysis of Pseudomonas aeruginosa quorum-sensing regulons: effects of growth phase and environment. J Bacteriol 185:2080–2095. https://doi.org/10.1128/JB.185.7.2080-2095.2003

Part VI

Culture-Based Methods

Chapter 17

Culture-Based Quantification of Biofilms Using Swab Sampling and CFU Enumeration

Sujith P

Abstract

Accurate quantification of viable cells within surface-associated biofilms is essential for understanding microbial adhesion, persistence, and antimicrobial responses. In this study, a standardized culture-based protocol was employed for biofilm sampling, recovery, and enumeration from defined solid surfaces. Biofilms formed on substrates such as glass slides, coupons, or quartz crystal microbalance (QCM) crystals were aseptically recovered using systematic swabbing with phosphate-buffered saline (PBS). Adhered cells were detached through vortexing and low-power sonication to ensure effective dispersion of biofilm aggregates while maintaining cell viability. Recovered suspensions were serially diluted and plated on nutrient and biofilm-supporting agar media, followed by incubation under appropriate conditions. Viable cell counts were determined by colony-forming unit (CFU) enumeration and expressed as CFU per milliliter or normalized to surface area (CFU/cm^2) for comparative analysis. The method is adaptable for single-species and mixed-species biofilms and allows optional enrichment or selective plating for targeted microbial populations. This reproducible and scalable approach provides a robust framework for quantitative biofilm analysis in environmental, industrial, and biomedical microbiology applications.

Key words Biofilm recovery, Surface-associated microorganisms, Swab sampling, Viable cell enumeration, Colony-forming units (CFU), Serial dilution plating, Quartz crystal microbalance (QCM), Microbial adhesion

1 Introduction

Culture-based methods are fundamental techniques in microbial biofilm research, providing a direct and reliable approach to quantify viable and culturable cells within surface-associated microbial communities. Unlike advanced molecular or imaging-based techniques, which focus on structural organization or total biomass, culture-based assays specifically target the actively growing and metabolically functional fraction of biofilm populations [1].

These methods such as swabbing of biofilm surfaces, culturing on appropriate agar media, serial dilution, and colony-forming unit (CFU) enumeration are simple, cost-effective, and highly

Abirami K (ed.), *Microbial Biofilm Research: Comprehensive Protocols and Best Practices*, Springer Protocols Handbooks, https://doi.org/10.1007/978-1-0716-5535-1_17,

reproducible. They allow researchers to assess the effectiveness of antimicrobial treatments, compare biofilm development under different environmental conditions, and monitor surface contamination in clinical, industrial, and environmental settings [2].

In biofilm research, culture-based approaches are often integrated with complementary techniques like microscopy (e.g., CLSM, SEM), spectroscopy, and molecular methods (e.g., qPCR, metagenomics) to provide a comprehensive understanding of biofilm formation, viability, and resistance characteristics. One limitation, however, is that culture-based methods may underestimate total cell numbers because they do not detect viable but non-culturable (VBNC) cells, which are known to persist in mature biofilms and contribute to their resilience [3, 4].

Despite this limitation, CFU-based quantification remains the gold standard for determining biofilm viability and microbial load, especially in studies focused on culturable fractions, treatment efficacy, and environmental persistence.

The general approach involves:

1. Swabbing or detaching biofilm cells from a surface.
2. Suspending and dispersing them in buffer.
3. Performing serial dilutions.
4. Plating on agar.
5. Calculating colony forming units (CFU).

These methods are crucial for evaluating biofilm formation kinetics, disinfectant efficacy, antibiotic tolerance, and surface-material interactions. They can be applied to static (e.g., QCM surfaces) or dynamic (e.g., flow cell) biofilm systems [5–7].

2 Materials

2.1 Sampling and Biofilm Recovery

- Sterile cotton or polyester swabs (or sterile cell scrapers/pipette tips).
- Sterile phosphate-buffered saline (PBS) or saline buffer.
- Sterile tubes (1.5 mL or 15 mL conical).
- Sonication bath or vortex mixer.
- Sterile templates or pre-defined surface area for swabbing.

2.2 Culturing and Dilution

- Nutrient agar or biofilm-specific media (e.g., LB, R2A, tryptic soy agar).
- Nutrient broth or equivalent enrichment media.
- Sterile dilution tubes.

- Micropipettes and sterile tips.
- Vortex mixer.

2.3 Plating and Enumeration

- Petri dishes.
- Glass or disposable spreaders.
- Incubator (temperature depending on microorganism).
- Marker pens and labels.
- Colony counter/magnifying lens.

2.4 Safety and Waste

- Sterile gloves and PPE.
- Biohazard bags.
- Autoclave/disinfectant solutions.

3 Methods

3.1 Biofilm Sampling/Swabbing

1. If biofilm is grown on a defined surface (e.g., QCM crystal, coupon, glass slide), gently remove the substrate using sterile forceps.
2. Using a sterile swab moistened with PBS, swab the entire surface systematically in horizontal and vertical strokes.
3. Alternatively, if working with a flow cell system, flush the chamber with a defined volume (e.g., 5 mL PBS) to detach loosely adhered cells, then swab or scrape remaining cells.
4. Insert the swab into a sterile tube containing 1–5 mL PBS buffer.
5. Label the sample with relevant information (ID, time point, surface type) [8] (*see* **Notes 1, 6** and **7** and Fig. 1).

3.2 Cell Recovery from Swab/Surface

1. Vortex the tube with the swab for 1 min to dislodge the cells into the buffer [8].
2. Optionally, sonicate the sample for 5 min in a low-power ultrasonic bath to disperse biofilm aggregates without killing cells.
3. Vortex again for 30 s.
4. Remove the swab aseptically and discard it in biohazard waste (*see* **Notes 2** and **8**).

3.3 Serial Dilution

1. Prepare serial dilution tubes with 900 μL PBS.
2. Transfer 100 μL of biofilm suspension into the first dilution (10^{-1}) and mix thoroughly.
3. Continue serial dilution to 10^{-6} or as required.
4. Use fresh tips for each dilution step (*see* **Notes 3** and **9**).

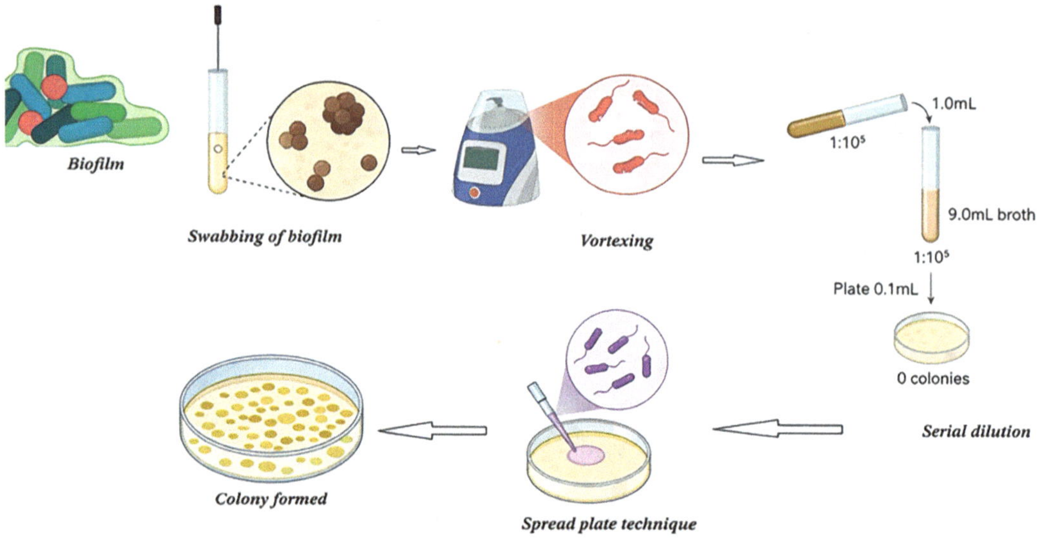

Fig. 1 Enumeration of biofilm-associated bacteria using serial dilution and spread plate technique. Biofilm samples were collected by sterile swabbing and transferred into a test tube containing broth. The sample was vortexed thoroughly to dislodge and homogenize bacterial cells. A series of tenfold serial dilutions (e.g., 10^{-5} to 10^{-6}) was prepared by transferring 1.0 mL of the sample into 9.0 mL of sterile broth. An aliquot (0.1 mL) from the appropriate dilution was plated onto agar plates using the spread plate technique. After incubation, visible colonies were formed and counted to determine the viable bacterial load in the original biofilm sample

3.4 Plating and Incubation

1. Plate 100 μL from selected dilutions (e.g., 10^{-2} to 10^{-6}) onto nutrient agar plates.
2. Spread evenly using sterile spreaders.
3. Let the plates absorb excess liquid.
4. Incubate inverted plates at the appropriate temperature for 24–48 h (*see* **Notes 4** and **10**).

3.5 CFU Counting

1. Select plates with 30–300 colonies.
2. Count colonies and record dilution factors.
3. Calculate viable cell numbers using the formula (*see* **Notes 5** and **11**).

$$\text{CFU/ml} = \frac{\text{Noof colonies} \times \text{dilution factor}}{\text{Volume plated (ml)}}$$

4. If biofilm surface area was defined, express as

$$\text{CFU/cm}^2 = \frac{\text{CFU recovered}}{\text{Surface area (cm}^2\text{)}}$$

3.6 Enrichment or Differential Plating

1. For mixed-species biofilms or low biomass, enrichment may be performed prior to dilution (*see* **Note 12**) [9].
2. Selective or differential media (e.g., MacConkey, Cetrimide) can be used to target specific populations.

4 Notes

1. Swabbing should be standardized (same pressure, area, and strokes) to ensure reproducibility.
2. Mild sonication improves recovery of cells embedded in EPS without significant loss of viability.
3. Proper mixing at each dilution is critical for reliable CFU quantification.
4. Avoid plating from undiluted biofilm samples, as biofilm suspensions often have high cell densities.
5. Counting at least three replicates improves accuracy.
6. Detachment efficiency varies with biofilm maturity and surface type; therefore, the protocol should be standardized for comparative studies.
7. Gentle sonication is effective for dispersing biofilms without killing cells.
8. Avoid excessive vortexing or harsh treatment, which can lyse cells.
9. For flow cell systems, detachment can also be achieved by flushing with buffer and backflow.
10. Record all dilution factors meticulously.
11. Plates with confluent growth should be excluded from CFU calculation.
12. Work under aseptic conditions to avoid contamination.

Competing Interest

The author(s) report no financial conflicts related to the subject matter of this chapter.

References

1. Stewart PS, Franklin MJ (2008) Physiological heterogeneity in biofilms. Nat Rev Microbiol 6(3):199–210. https://doi.org/10.1038/nrmicro1838
2. Hall-Stoodley L, Costerton JW, Stoodley P (2004) Bacterial biofilms: from the natural environment to infectious diseases. Nat Rev Microbiol 2(2):95–108. https://doi.org/10.1038/nrmicro821
3. Oliver JD (2010) Recent findings on the viable but nonculturable state in pathogenic bacteria.

FEMS Microbiol Rev 34(4):415–425. https://doi.org/10.1111/j.1574-6976.2009.00200.x
4. Li L, Mendis N, Trigui H, Oliver JD, Faucher SP (2014) The importance of the viable but non-culturable state in human bacterial pathogens. Front Microbiol 5:258. https://doi.org/10.3389/fmicb.2014.00258
5. Bridier A, Briandet R, Thomas V, Dubois-Brissonnet F (2015) Resistance of bacterial biofilms to disinfectants: a review. Biofouling 27(9):1017–1032. https://doi.org/10.1080/08927014.2011.626899
6. Flemming H-C, Wingender J, Szewzyk U, Steinberg P, Rice SA, Kjelleberg S (2016) Biofilms: an emergent form of bacterial life. Nat Rev Microbiol 14(9):563–575. https://doi.org/10.1038/nrmicro.2016.94
7. Sutton S (2011) Measurement of microbial cells by colony-forming units. Journal of Validation Technology 17(3):42–46
8. International Organization for Standardization (ISO) (2018) ISO 18593:2018. Microbiology of the food chain—Horizontal methods for surface sampling. Geneva, Switzerland: ISO
9. ASTM International (2013) ASTM E2871-13: standard test method for evaluating disinfectant efficacy against biofilm using the CDC biofilm reactor. West Conshohocken, PA: ASTM International https://doi.org/10.1520/E2871-13

Part VII

Spectroscopic and Chemical Analysis

Chapter 18

Rapid Detection of Biofilms Using Fourier Transform Infrared Spectroscopy

Prabha Nagaram

Abstract

Bacterial biofilms are associated with many persistent infections because once they attach to surfaces and produce their matrix, they form highly resistant communities compared to free-living (planktonic) bacteria. The cells within a biofilm are embedded in an extracellular polymeric substance (EPS), which consists of polysaccharides, proteins, lipids, and nucleic acids. This matrix provides protection, promotes survival and gene exchange, and shields the bacteria from antimicrobial treatments. Therefore, understanding how biofilms develop and how these sessile cells endure hostile conditions is essential for designing new therapeutic strategies. Conventional morphological and biochemical techniques for identifying microorganisms are often labor-intensive, slow, and expensive. In contrast, vibrational spectroscopic approaches such as FTIR spectroscopy offer a faster and more precise option. FTIR provides a "whole-organism fingerprint" by detecting the biochemical composition of cells, improving screening processes especially in clinical microbiology for more accurate identification of infection sources. Additionally, its capability to examine living samples without disruption over extended periods makes it particularly useful for studying biofilm development. These limitations of traditional methods emphasize the need for advanced technologies to support the diagnosis, characterization, and treatment of biofilm-related infections. Among the promising new tools, Fourier Transform Infrared (FTIR) spectroscopy has emerged as a valuable technique for biofilm analysis.

Key words Biofilm, FTIR, Lyophilized cells, Chronic infection

1 Introduction

Biofilms are considered the predominant lifestyle of bacteria and have crucial implications in industrial, clinical, and agricultural settings. They consist of microbial communities that permanently attach to living or non-living surfaces and are embedded within an extracellular polymeric substance (EPS) made up of polysaccharides, nucleic acids, proteins, and lipids [1]. These biofilm-associated bacteria pose a major challenge to public health and contribute significantly to infection-related diseases. The National Institutes of Health estimates that approximately 65% of all microbial

Abirami K (ed.), *Microbial Biofilm Research: Comprehensive Protocols and Best Practices*, Springer Protocols Handbooks,
https://doi.org/10.1007/978-1-0716-5535-1_18,

infections, and nearly 80% of persistent wound infections, are associated with biofilm formation [2].

To improve treatment outcomes, it is crucial for medical microbiologists to precisely detect and characterize biofilm-associated infections using clinically meaningful indicators, as this supports the development of complementary therapeutic strategies that can minimize the required antibiotic dosage. Traditional diagnostic approaches typically assess phenotypic and genotypic traits—such as species, strain type, virulence factors, antibiotic resistance profiles, and metabolic characteristics—but their limited sensitivity often leads to the need for multiple tissue biopsies from areas surrounding the suspected infection site (e.g., around medical implants) [3]. When biofilm infections are confirmed, they may necessitate surgical cleaning or removal of the infected implant. Due to these challenges, including high cost, lengthy processing time, and lack of standardized procedures, there is a pressing need for innovative and reliable methods for biofilm detection [4].

Analyzing biofilms is often difficult due to their complex structure, diverse composition, and variability. Fourier transform infrared (FTIR) spectroscopy has emerged as a promising analytical approach, offering a rapid, low-cost, objective, and dependable means for clinically assessing biofilm-related infections [5]. Unlike traditional metabolic assays that require time-consuming culturing, FTIR provides quicker results and remains more affordable and accessible compared to newer molecular diagnostic techniques. Notably, studies have demonstrated that the ability of FTIR to differentiate between microbial strains is comparable to that of whole-genome sequencing (WGS), which is widely regarded as the benchmark for genotypic analysis [6].

Alongside taxonomic and molecular approaches, FTIR spectroscopy offers distinctive fingerprint spectra that enable rapid, non-destructive identification of microbial species. Moreover, it allows for prolonged, viable, and undisturbed monitoring of samples, making it particularly valuable for foundational studies on biofilm development and behavior [7].

FTIR spectroscopy is a chemical analysis method that records absorbance spectra reflecting the functional groups present in biological samples, thereby representing their overall molecular composition. A key benefit of FTIR is its flexibility, as it allows samples to be analyzed using various measurement configurations and growth surfaces, which can streamline the analytical process [8]. However, due to the structural complexity and variability of microbial biofilms, deciphering their behavior remains challenging without improved monitoring technologies. Consequently, there is a growing need for methods capable of providing accurate insights into the diverse chemical and physiological conditions that exist within biofilms.

FTIR spectroscopy provides a fast and non-destructive method for microbial characterization, requiring only small sample volumes while still delivering strong signal quality. When combined with Principal Component Analysis (PCA), it can reliably differentiate between microbial strains. However, some limitations exist, including interference from water absorption, overlapping spectral bands from unrelated functional groups, variability caused by differences in growth conditions, and the inability to detect simple diatomic molecules such as O_2 and N_2. Nonetheless, in recent years, significant advancements have been made in FTIR instrumentation and associated optical components [9].

2 Materials

Microbial samples are typically analyzed either as liquid suspensions in a flow cell or as freeze-dried biofilm material that is combined with potassium bromide (KBr) and pressed into a pellet for assessment using attenuated total reflectance (ATR)-FTIR.

- Lyophilized (freeze-dried) bacterial cells are required as the sample.
- Spectroscopy-grade potassium bromide (KBr) is required.
- A hot air oven and a desiccator are used for drying the samples.
- A pellet sample holder is required.
- A pellet die kit suitable for preparing KBr discs is required, although a simple mortar and pestle may also be used depending on the nature of the sample.
- A rotary vacuum pump is needed.
- A hydraulic press (10–12 tons pressure) is needed.

3 Methods

3.1 Steps for Preparing Potassium Bromide Pellets

KBr pellets are transparent to infrared light, making them suitable for preparing samples for FTIR measurements.

1. Finely grind the KBr salt and dry it in a hot air oven at 120 °C for a minimum of 5 h, then store it in a desiccator to prevent moisture absorption (*see* **Note 1**).
2. Mix approximately 1 mg of the sample with 100 mg of KBr and gently grind them together using a mortar and pestle. Avoid excessive grinding, as KBr is hygroscopic and must be protected from moisture.
3. Place a polished die pellet into the pellet die kit with the smooth side facing upward. Carefully add the prepared

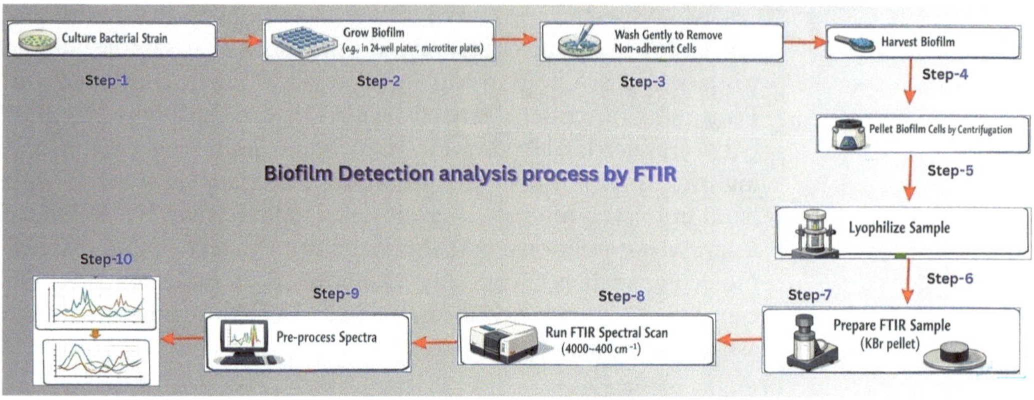

Fig. 1 Biofilm detection and compositional analysis by Fourier transform infrared (FTIR) spectroscopy. This figure illustrates the workflow and representative spectral output for the detection and biochemical characterization of biofilms using FTIR spectroscopy

KBr–sample mixture into the die cavity and use the plunger to spread it evenly (*see* **Note 2**).

4. Next, position the second die pellet on top, polished side facing downward, using the plunger to guide it into place (*see* **Note 3**).
5. Position the die assembly containing the sample into the hydraulic press, and then use a rotary pump to create a vacuum inside the chamber.
6. While maintaining the vacuum, apply a pressure of approximately 8–10 tons for about 5 min using the hydraulic press. Take care to avoid cracking the pellet. For die sizes between 10 and 13 mm, this pressure range is typically sufficient.
7. Slowly release the applied pressure by turning off the vacuum. Avoid releasing the pressure too quickly, as this may cause the pellet to fracture.
8. Gently remove the die assembly from the hydraulic press and detach the base. Then invert the die and carefully push from the upper section to release the lower pellet without applying excessive force.
9. For FTIR measurement, gently remove the formed KBr pellet from between the die plates using a spatula and place it into the sample holder. Take care not to scratch the smooth surfaces of the pellet (*see* **Note 4**).
10. The spectrum should be recorded immediately to minimize moisture uptake from the surrounding air.
11. All components, including the die chamber, disc pellets, and plunger, should be thoroughly cleaned using a soft tissue lightly moistened before storage or reuse (*see* **Notes 5** and **6**) (Fig. 1).

Culture Bacterial Strain

↓

Grow Biofilm
(e.g., in 24-well plates, microtiter plates)

↓

Wash Gently to Remove
Non-adherent Cells

↓

Harvest Biofilm

↓

Pellet Biofilm Cells by Centrifugation

↓

Lyophilize Sample

↓

Prepare FTIR Sample
(KBr pellet)

↓

Run FTIR Spectral Scan
(4000–400 cm^{-1})

↓

Pre-process Spectra

↓

Analyse Spectral Data

The flowchart represents stepwise procedure for analysis of biofilm-forming organisms by FTIR.

4 Notes

1. If the pellets appear cloudy or brittle, it is likely due to moisture absorbed by the KBr or the sample. To prevent this, ensure the KBr is well dried, limit its exposure to air, and reduce the duration of grinding.
2. If the pellets crack while being pressed, it is likely due to applying too much pressure too quickly. To prevent this, increase the pressure gradually, keep it within the recommended 8–10-ton range, and make sure the sample and KBr are completely dry.
3. If the absorption peaks appear unusually strong, it may indicate that too much sample was used. To correct this, decrease the sample quantity to about 0.5–1 mg.
4. During FTIR sample preparation, the lyophilization of bacterial cells must be carried out carefully. Additionally, avoid

applying excessive force with the plunger, as this can cause the KBr to adhere to the surfaces.

5. For biofilm samples of clinical origin, flow rates in the range of 3–5 mL/h are commonly used, while biofilms in drinking water systems are typically examined at higher flow rates, around 45 mL/h.
6. The length of data acquisition varies depending on the type of experiment and the bacterial growth period. To achieve a high signal-to-noise ratio, multiple repeated scans are recommended, provided that no significant spectral variations occur during the measurement.

Competing Interest

The author declares that they have no competing interests. All aspects of the work were conducted independently, without any financial, personal, or professional relationships that could be perceived as influencing the results, interpretation, or presentation of this study.

References

1. Costerton JW, Lewandowski Z, Caldwell DE, Korber DR, Lappin-Scott HM (1995) Microbial biofilms. Ann Rev Microbiol 49:711–745. https://doi.org/10.1146/annurev.mi.49.100195.003431
2. Jamal M, Ahmad W, Andleeb S, Jalil F, Imran M, Nawaz MA, Hussain T, Ali M, Rafiq M, Kamil MA (2018) Bacterial biofilm and associated infections. J Chin Med Assoc 81(1):7–11. https://doi.org/10.1016/j.jcma.2017.07.012
3. Wu H, Moser C, Wang HZ, Høiby N, Song ZJ (2015) Strategies for combating bacterial biofilm infections. Int J Oral Sci 7(1):1–7. https://doi.org/10.1038/ijos.2014.65
4. Winkler H (2017) Treatment of chronic orthopaedic infection. EFORT Open Rev 2(5):110–116. https://doi.org/10.1302/2058-5241.2.160063
5. Naumann D (2008) Vibrational spectroscopy in microbiology and medical diagnostics. In: Lasch P, Kneipp J (eds) Biomedical vibrational spectroscopy. John Wiley & Sons, Hoboken, NJ, pp 1–8
6. Novais A, Freitas A, Rodrigues C, Peixe L (2019) Fourier transform infrared spectroscopy: unlocking fundamentals and prospects for bacterial strain typing. Eur J Clin Microbiol Infect Dis 38:427–448. https://doi.org/10.1007/s10096-018-3431-3
7. Schmitt J, Flemming HC (1998) FTIR-spectroscopy in microbial and material analysis. Int Biodeterior Biodegradation 41(1):1–11. https://doi.org/10.1016/S0964-8305(97)00086-0
8. Alvarez-Ordóñez A, Mouwen DJM, López M, Prieto M (2011) Fourier transform infrared spectroscopy as a tool to characterize molecular composition and stress response in foodborne pathogenic bacteria. J Microbiol Methods 84(3):369–378. https://doi.org/10.1016/j.mimet.2011.01.005
9. Kamnev AA, Dyatlova YA, Kenzhegulov OA, Fedonenko YP, Evstigneeva SS, Tugarova AV (2023) Fourier transform infrared (FTIR) spectroscopic study of biofilms formed by the rhizobacterium *Azospirillum baldaniorum* Sp245: aspects of methodology and matrix composition. Molecules 28(4):1949. https://doi.org/10.3390/molecules28041949

Chapter 19

A Stepwise Approach to Biofilm Detection Using Nuclear Magnetic Resonance

Abirami K

Abstract

Biofilms are structured microbial communities encased in an extracellular polymeric matrix that provide protection to microorganisms and contribute to persistent infections and industrial fouling. Reliable detection and characterization of biofilm formation remain essential for understanding their development and resistance behavior. In this study, nuclear magnetic resonance (NMR)-based techniques were employed as noninvasive tools for detecting and monitoring biofilm growth. Proton (^{1}H) NMR relaxometry and spectral profiling approaches were used to assess changes in water mobility, matrix density, and metabolic composition during different stages of biofilm formation. Progressive alterations in transverse relaxation times (T_2) indicated increasing structural complexity and restricted molecular motion within the developing matrix. Complementary spectral analysis revealed shifts in metabolite patterns associated with extracellular polymer production and cellular adaptation. The results demonstrate that NMR provides quantitative, reproducible, and label-free detection of biofilms without disturbing their native structure. This approach offers significant potential for real-time monitoring of biofilm development in biomedical, environmental, and industrial systems.

Key words Biofilm detection, Nuclear Magnetic Resonance (NMR), ^{1}H NMR spectroscopy, NMR relaxometry, Magnetic Resonance Imaging (MRI), Pulsed Field Gradient NMR (PFG-NMR), Diffusion analysis, Noninvasive monitoring

1 Introduction

Nuclear magnetic resonance (NMR) is a powerful analytical technique that detects the magnetic properties of atomic nuclei most commonly hydrogen (^{1}H) or carbon (^{13}C)—when they interact with an external magnetic field and radiofrequency pulses. The resulting signals provide detailed information about molecular structure, dynamics, and environment. The biofilms surface attached microbial communities encased in extracellular polymeric substances (EPS) NMR offers a noninvasive, label-free way to probe internal chemistry, hydration, and structural changes without destroying the biofilm. This is particularly useful because

Abirami K (ed.), *Microbial Biofilm Research: Comprehensive Protocols and Best Practices*, Springer Protocols Handbooks,
https://doi.org/10.1007/978-1-0716-5535-1_19,

biofilms are opaque and complex, making many optical methods difficult to apply [1].

One common approach for detecting biofilm growth, especially in porous or environmental systems, relies on measuring NMR relaxation times, particularly spin–spin relaxation (T_2). As biofilms form within a medium (e.g., sand, soil, porous materials), they alter the local water environment and restrict molecular mobility. These changes shorten T_2 values and shift relaxation distributions—patterns that can be monitored over time. Low-field, portable NMR instruments have been used successfully to track biofilm growth in real time by detecting these changes. For example, changes in T_2 relaxation over days corresponded directly with increasing biofilm presence in laboratory porous media models [2].

Beyond simple detection, high-resolution NMR and solid-state NMR allow researchers to dig deeper into the biochemical makeup of biofilms. These methods can:

- Identify and quantify specific metabolites within biofilm cells and EPS.
- Distinguish between different chemical environments of biofilm components.
- Reveal structural features of macromolecules such as polysaccharides and proteins.

A recent study using dynamic nuclear polarization-enhanced solid-state NMR (DNP ssNMR) dramatically improved sensitivity by ~75-fold compared to conventional solid-state NMR permitting detailed analysis of biofilm composition without isotope labeling. This advance allows researchers to observe subtle chemical features of native biofilms on realistic time scales and without extensive sample manipulation [3].

NMR is particularly powerful at revealing the chemical fingerprints of metabolites and structural molecules inside biofilms. For example, NMR metabolomics can distinguish differences in metabolic pathways depending on growth conditions, a capability recently demonstrated in *Burkholderia pseudomallei* biofilms, where metabolites like acetate and betaine showed distinct patterns depending on nutrient conditions. Solid-state NMR has been used to quantify the relative abundance of polysaccharides, proteins, and other macromolecules in the EPS matrix information that is crucial to understanding the physical resilience and diffusion barriers of biofilms. These insights can inform strategies for disrupting or degrading biofilms in clinical or industrial settings [4]. Despite its strengths, NMR has limitations. Low-field instruments have limited spatial resolution, making it difficult to visualize fine biofilm structures directly. High-field, high-resolution techniques require sophisticated equipment and expertise, which may not be accessible in every laboratory. However, ongoing advances like DNP enhancement and integrated NMR imaging techniques are pushing the boundaries of what's possible, enabling richer, faster, and more detailed biofilm analyses.

NMR is emerging as not just a tool for detecting biofilms but as a comprehensive analytical platform capable of revealing chemical, structural, and dynamic features of biofilm communities. From tracking growth non-invasively to decoding the biochemical makeup of EPS and cellular metabolism, modern NMR methods are helping scientists better understand and eventually control biofilms in medicine, environment, and industry.

2 Materials

2.1 Nuclear Magnetic Resonance Spectrometer

- High-field system (e.g., 400–800 MHz) for detailed spectral analysis.
- Low-field bench-top NMR for relaxometry studies.
- Must support ^{1}H detection (and optionally ^{13}C or ^{31}P).

NMR Probe

- Liquid-state probe (for metabolomics or EPS extracts).
- Solid-state probe with MAS (Magic Angle Spinning) for intact biofilms.
- Diffusion probe (for PFG/DOSY experiments).

Gradient System

- Required for pulsed field gradient experiments (diffusion measurements).

Temperature Control Unit

- Maintains stable temperature (usually 25–37 °C for biological samples).

2.2 Biofilm Cultivation Materials

Microbial Strain

- Known biofilm-forming organism (e.g., *Pseudomonas*, *Staphylococcus*).

Growth Media

- LB broth, TSB, or minimal medium.
- Sterile buffers (e.g., PBS).

Biofilm Growth Surfaces

- Glass slides.
- Polystyrene plates.

- Stainless steel coupons.
- Microfluidic flow cells (if dynamic biofilm is studied).

Incubator

- Controlled temperature and humidity.

2.3 Sample Preparation Materials

NMR Tubes

- 5 mm standard tubes (liquid NMR).
- MAS rotors (solid-state NMR).

Deuterated Solvents

- D_2O (deuterium oxide).
- Deuterated buffer systems (to reduce solvent proton signal).

Internal Standards

- TMS (tetramethylsilane).
- DSS (sodium trimethylsilylpropionate).

Centrifuge

- For separating planktonic cells from biofilm matrix.

Lyophilizer (Optional)

- For drying samples before solid-state NMR.

2.4 Chemicals and Reagents

- Phosphate-buffered saline (PBS).
- Sterile distilled water.
- Isotope-labeled substrates (optional, e.g., ^{13}C-glucose).
- EPS extraction reagents (e.g., EDTA, NaCl solutions).

2.5 Data Acquisition and Analysis Tools

NMR Software

- Bruker TopSpin.
- MNova.
- MestReNova.
- MATLAB or SIMCA (for multivariate analysis).

Computer Workstation

- High processing capability for 2D spectra and MRI reconstruction.

2.6 Validation and Supporting Equipment

- Confocal microscope (for structural comparison).
- Spectrophotometer (biofilm biomass estimation).

- Autoclave (for sterilization).
- Analytical balance.

2.7 Safety and Laboratory Supplies

- Gloves, lab coat, safety goggles.
- Magnetic-safe tools (non-ferromagnetic tweezers).
- Cryogens (liquid nitrogen/helium, if required for high-field systems).

3 Methods

3.1 Prepare Biofilm Samples

Proper sample prep is foundational:

(a) Grow or Harvest Biofilm

- Cultivate the biofilm on defined surfaces (glass, beads, microchannel, etc.) until mature.
- Rinse gently to remove planktonic cells while preserving the matrix [5].

(b) Transfer to NMR-Compatible Holder

- Place intact biofilm or collected biomass into an NMR tube or microfluidic sample holder.
- For liquid-state NMR: cells often need to be suspended in deuterated buffer to minimize solvent signal.
- For solid or gel biofilms: use solid-state NMR rotors or gel-packed probes.

Tip: Recent metabolomics work using ^{1}H NMR highlights the need for careful handling to preserve metabolite profiles.

3.1.1 Acquire the NMR Data

(a) Calibration and Setup

- Tune the probe to the resonance frequency of ^{1}H (often ~400–800 MHz in high-field spectrometers).
- Use standard samples (like TMS or DSS) for chemical shift referencing.
- Optimize shimming (magnetic field homogeneity) for best resolution.

(b) Run Experiments

- Relaxometry: Use a CPMG or inversion-recovery pulse sequence to measure T1/T2 decays (*see* **Notes 1** and **2**).
- Diffusion (PFG): Use gradient pulses to quantify diffusion coefficients.

- 1D/2D Spectroscopy: Collect ^1H or two-dimensional spectra for metabolite/cross-peak analysis.

(c) Controls

- Include a non-biofilm control (e.g., buffer or planktonic cells) to differentiate signals.
- Run replicate measurements for statistical reliability.

3.1.2 Process and Analyze the Data

(a) Transform Raw Signals

- Use Fourier transform to convert time-domain signals to spectra.
- Phase and baseline correct as needed.

(b) Interpret Relaxation and Diffusion

- Compare T1/T2 distributions—shorter relaxation suggests more bound water in biofilm matrix.
- Diffusion coefficients lower than free water indicate restriction due to dense biofilm.

(c) Identify Chemical Features

- Metabolite peaks or EPS signals (e.g., carbohydrates, amino resolvants) help confirm biofilm biochemical traits.
- Multivariate analyses (PCA, clustering) can differentiate biofilm vs non-biofilm profiles.

3.1.3 Validate and Correlate Findings

NMR results gain strength when confirmed with complementary techniques:

- Microscopy (e.g., confocal) to visualize structure.
- Mass spectrometry to confirm metabolite identities.
- Biofilm growth assays to correlate physiology with NMR signals.

This integrated approach ensures NMR findings reflect true biofilm characteristics rather than artifacts.

3.2 Detection of Biofilm Presence and Growth Using NMR Relaxometry (T_1/T_2)

1. Biofilm formation and growth were monitored non-destructively using proton (^1H) nuclear magnetic resonance (NMR) relaxometry. Longitudinal (T_1) and transverse (T_2) relaxation times were measured at defined incubation intervals to evaluate changes in water mobility and matrix density associated with biofilm development. All experiments were performed in triplicate to ensure reproducibility.

2. Relaxation measurements were conducted using a bench-top or high-field nuclear magnetic resonance spectrometer operating at a proton frequency appropriate to the instrument configuration (typically 20–60 MHz for low-field systems or 400–600 MHz for high-field systems). The instrument was equipped with a ^{1}H detection probe and temperature control unit. All measurements were performed at a constant temperature of 25 ± 0.5 °C (or 37 °C when physiological conditions were required).
3. Prior to data acquisition, probe tuning, matching, and magnetic field shimming were optimized to ensure signal stability and homogeneity.
4. A biofilm-forming bacterial strain (e.g., *Pseudomonas aeruginosa*) was cultured overnight in sterile growth medium (Luria–Bertani broth or tryptic soy broth) at 37 °C with agitation (150 rpm). The overnight culture was diluted 1:100 in fresh medium and dispensed into sterile glass tubes or onto sterile coupons placed within tubes.
5. Biofilms were grown under static conditions at 37 °C for 0, 12, 24, 48, and 72 h. At each time point, non-adherent planktonic cells were removed by gently washing twice with sterile phosphate-buffered saline (PBS), taking care not to disturb attached biofilm structures.
6. Two preparation approaches were used depending on the experimental objective:
   - Biofilm-coated surfaces were carefully transferred into 5 mm NMR tubes. A minimal volume of sterile PBS was added to maintain hydration. Air bubbles were avoided to prevent magnetic susceptibility artifacts.
   - Biofilm biomass was gently scraped, resuspended in a fixed volume of PBS, and transferred to NMR tubes. All samples were standardized to identical volumes to ensure consistency in relaxation measurements.
   - Control samples included sterile medium and planktonic cell suspensions.
7. Transverse relaxation times (T_2) were measured using the Carr–Purcell–Meiboom–Gill (CPMG) pulse sequence (*see* **Note 3**).
   - *Acquisition Parameters:*

     Repetition time (TR): 3–5 × estimated T_1

     Echo time (TE): 0.2–1.0 ms

     Number of echoes: 1000–5000

     Number of scans: 8–32

8. The decay of transverse magnetization was recorded and stored as a multi-exponential relaxation curve. Longitudinal relaxation times (T_1) were determined using an inversion-recovery pulse sequence.
   - *Acquisition Parameters:*
     180° inversion pulse
     Variable delay times (10–15 increments)
     90° detection pulse

   Repetition delays sufficient for full relaxation. Signal intensities were recorded as a function of inversion delay and fitted to an exponential recovery model.
9. Relaxation measurements were performed at predetermined incubation intervals (0–72 h). T_2 values were plotted as a function of incubation time to assess biofilm maturation. A progressive reduction in T_2 values was interpreted as increased matrix density and restricted water mobility within the developing biofilm.
10. T_1 measurements were used as complementary indicators of changes in molecular interactions between water and extracellular polymeric substances (EPS).
11. Raw free induction decay (FID) signals were processed using the instrument's proprietary software. Exponential fitting was applied to determine T_1 and T_2 relaxation constants.
12. Where applicable, multi-exponential analysis or inverse Laplace transformation was performed to resolve distinct water populations (free vs. bound water fractions). Statistical analysis was conducted using triplicate measurements, and significance was determined using one-way ANOVA with a confidence level of $p < 0.05$.
13. To corroborate NMR findings, biofilm biomass was quantified using complementary methods such as crystal violet staining or optical density measurements. Structural verification was performed, when necessary, using microscopic analysis.
14. All experiments were conducted in triplicate under identical environmental conditions. Control measurements from sterile medium and planktonic cultures were used to establish baseline relaxation values. Instrument calibration was verified daily prior to experimental measurements.
15. Expected Outcome: Biofilm formation was characterized by a progressive decrease in T_2 relaxation times, reflecting restricted molecular mobility and increasing matrix density. T_1 relaxation times exhibited moderate variation corresponding to changes in hydration state and macromolecular interactions. Together, these parameters provided a quantitative and non-destructive measure of biofilm presence and growth kinetics (*see* **Notes 4** and **5**).

3.3 Biofilm Detection by Pulsed Field Gradient NMR (PFG-NMR)

Pulsed field gradient nuclear magnetic resonance (PFG-NMR) measures molecular self-diffusion by encoding spatial position into the phase of nuclear spins using magnetic field gradients. In biofilms, water and small molecules experience restricted diffusion due to the dense extracellular polymeric substance (EPS) matrix. As biofilm biomass increases, diffusion coefficients decrease relative to planktonic or free-water controls. Thus, monitoring changes in apparent diffusion coefficients (D) provides a quantitative and non-destructive indicator of biofilm presence, structural development, and matrix density.

3.3.1 Instrumentation

Experiments were performed using a nuclear magnetic resonance spectrometer equipped with:

- ^{1}H detection probe
- Pulsed field gradient (PFG) system
- Temperature control unit
- Diffusion-compatible probe (recommended)

Typical operating frequencies:

- Low-field systems: 20–60 MHz
- High-field systems: 400–600 MHz

All measurements were conducted at controlled temperature (25 ± 0.5 °C or 37 °C).

3.3.2 Biofilm Cultivation

A biofilm-forming bacterial strain (e.g., *Pseudomonas aeruginosa*) was cultured overnight in sterile nutrient broth at 37 °C with agitation. The culture was diluted (1:100) into fresh medium and incubated under static conditions in sterile tubes or on glass/stainless steel coupons to allow surface attachment.

Biofilms were grown for defined intervals (0, 12, 24, 48, and 72 h). At each time point:

1. Planktonic cells were gently removed.
2. Samples were washed twice with sterile phosphate-buffered saline (PBS).
3. Care was taken to preserve structural integrity.

3.3.3 Sample Preparation

Two experimental configurations were used:

(a) Intact Biofilm Measurement

Biofilm-coated surfaces were placed inside 5 mm NMR tubes. Minimal PBS was added to maintain hydration. Air bubbles were carefully eliminated.

(b) Suspended Biofilm Matrix

Biofilm biomass was gently scraped and suspended in a fixed volume of PBS before transfer to NMR tubes.

Control samples included:

- Sterile medium
- Planktonic cell suspensions

Sample volumes were kept constant to ensure reproducibility.

3.3.4 PFG-NMR Diffusion Measurement

Diffusion measurements were performed using a stimulated echo or pulsed gradient spin echo (PGSE) sequence (*see* **Notes 6** and 7).

(a) Pulse Sequence Parameters

Typical parameters included:

- Gradient pulse duration (δ): 2–5 ms
- Diffusion delay (Δ): 20–100 ms
- Gradient strength (g): incremented linearly across 16–32 steps
- Repetition time (TR): $\geq 5 \times T_1$
- Number of scans: 8–32

Gradient amplitudes were varied systematically to obtain signal attenuation curves.

3.3.5 Data Acquisition and Processing

Signal attenuation was recorded as a function of gradient strength. Diffusion coefficients (D) were calculated using the Stejskal–Tanner equation:

$$I = I_0 \exp\left[-\gamma^2 g^2 \delta^2 \left(\Delta - \frac{\delta}{3}\right) D\right]$$

Where:

- I = observed signal intensity
- I_0 = initial signal
- γ = gyromagnetic ratio
- g = gradient strength
- δ = gradient duration
- Δ = diffusion delay
- D = diffusion coefficient

Data were fitted using mono- or multi-exponential models. In heterogeneous biofilms, multi-component diffusion behavior was expected, reflecting free water and matrix-bound water fractions.

3.3.6 Time-Course Monitoring of Biofilm Growth

Diffusion measurements were performed at sequential incubation times. The apparent diffusion coefficient (D_app) was plotted against growth time.

Expected trend:

Growth stage	Biofilm density	Diffusion coefficient
Initial (0 h)	No biofilm	High (≈ free water)
Early (24 h)	Thin layer	Moderate decrease
Mature (48 h)	Thick matrix	Significant reduction
Late (72 h)	Dense structure	Lowest D_app

A progressive decrease in D_app indicates increasing structural complexity and restricted molecular mobility.

3.3.7 Interpretation of Results

- Reduced diffusion confirms matrix formation.
- Multi-component diffusion suggests heterogeneous microenvironments.
- Strong restriction indicates mature biofilm development.
- Comparisons with planktonic controls differentiate true biofilm effects from cell density alone.

3.3.8 Validation and Complementary Analysis

To confirm PFG-NMR findings:

- Biomass quantification via crystal violet staining
- Microscopic imaging (e.g., confocal microscopy)
- Relaxometry (T_2) comparison for structural correlation

3.3.9 Statistical Analysis

All experiments were conducted in triplicate. Diffusion coefficients were reported as mean ± standard deviation. Statistical significance between time points was assessed using one-way ANOVA ($p < 0.05$).

Advantages of PFG-NMR for Biofilm Detection:

- Noninvasive and label-free
- Sensitive to structural heterogeneity
- Quantitative diffusion measurement
- Capable of real-time monitoring

3.4 Biofilm Analysis by ^1H NMR Spectral Profiling / Metabolomics

This method outlines a practical workflow for examining the metabolic composition of biofilms using proton (^1H) nuclear magnetic resonance (NMR) spectroscopy. Rather than targeting a single compound, this approach captures a broad metabolic "fingerprint" of the biofilm. It allows researchers to detect and quantify extracellular polymeric substances (EPS), intracellular metabolites, and biochemical changes that occur as the biofilm forms, matures, or responds to environmental stress [6].

By analyzing the ^{1}H NMR spectral profile at different growth stages or treatment conditions, it becomes possible to monitor shifts in metabolic pathways, nutrient utilization, and matrix production. Because NMR spectroscopy is highly reproducible and non-destructive, it provides reliable comparative data across experiments without the need for chemical labeling or extensive sample modification. Over all, ^{1}H NMR–based metabolomics offers a comprehensive and quantitative way to understand the biochemical dynamics underlying biofilm development.

3.4.1 Materials and Instrumentation

Instrumentation

- High-field nuclear magnetic resonance spectrometer (400–600 MHz recommended)
- ^{1}H detection probe
- Temperature control system (usually 25 °C)
- Centrifuge (≥10,000 rpm)
- Lyophilized (optional)

Reagents

- Deuterium oxide (D_2O)
- Phosphate buffer (prepared in D_2O, pH 7.0 ± 0.1)
- Internal standard (e.g., DSS–sodium trimethylsilyl propionate)
- Methanol (HPLC grade)
- Chloroform (optional for biphasic extraction)

Consumables

- 5 mm NMR tubes
- Microcentrifuge tubes
- Sterile PBS

3.4.2 Biofilm Cultivation

A biofilm-forming bacterial strain, such as *Pseudomonas aeruginosa*, is first introduced into sterile growth medium and incubated overnight at 37 °C with shaking to obtain an active starter culture. The following day, the culture is diluted (typically 1:100) into fresh medium and transferred to sterile tubes or appropriate growth surfaces. To encourage surface attachment and matrix development, the cultures are incubated under static conditions. Samples are collected at predetermined time points, commonly 24, 48, and 72 h, depending on the growth characteristics of the organism.

At each time point, non-adherent (planktonic) cells are carefully removed by gently washing the surface with sterile phosphate-buffered saline (PBS). This step ensures that subsequent analysis focuses on attached biofilm biomass rather than free-floating cells. All experiments should be performed in biological triplicates to ensure reliability and reproducibility of results.

3.4.3 Metabolite Extraction

Quenching of Metabolic Activity

To preserve the true metabolic state of the biofilm at the time of sampling, cellular metabolism must be halted immediately. The growth medium is removed quickly, and pre-chilled methanol (−20 °C) is added directly to the biofilm. Samples are kept on ice to prevent further metabolic reactions and degradation of metabolites. Rapid handling during this stage is essential to maintain accuracy in metabolic profiling (Table 1).

3.4.4 Extraction of Metabolites

(a) Polar Metabolite Extraction (Preferred for ^{1}H NMR Analysis).

For profiling water-soluble metabolites, a cold methanol–water mixture (2:1, v/v) is added to the quenched biofilm sample. The mixture is vortexed thoroughly to ensure efficient metabolite release. Sonication may be applied for 5–10 min to improve extraction efficiency. The sample is then centrifuged at 10,000 rpm for 10 min at 4 °C to separate cellular debris from soluble metabolites. The supernatant containing extracted metabolites is collected and dried either under a gentle nitrogen stream or by lyophilization (*see* **Note 8**).

(b) Methanol–Chloroform Extraction (For Broader Metabolite Coverage)

Table 1
Caption different types of NMR modes and its specific features

Goal	Typical NMR mode	Special features
Detect biofilm presence and growth	Relaxometry (T1/T2)	Measures how nuclear magnetization returns to equilibrium. Biofilm matrix often shows distinct relaxation times from free water due to restricted motion and higher biomass content
Map diffusivity and water dynamics	Pulsed field gradient NMR (PFG-NMR)	Applies magnetic gradients to track molecular diffusion— slower diffusion often correlates with dense biofilm regions
Chemical composition (metabolites/EPS)	^{1}H NMR spectral profiling / metabolomics	Produces spectra identifying metabolites and functional groups (such as sugars, amino acids, organic acids) that are characteristic of biofilm metabolism and EPS composition
Structural visualization	Magnetic resonance imaging (MRI)	Gives spatial maps of biofilm structure and water distribution, particularly valuable for thick or hydrated biofilms

Table summarizes how each NMR mode differs in terms of operational principles, key acquisition parameters, and analytical applications. Understanding these specifications ensures appropriate selection of experimental conditions based on the specific research objective, whether structural, dynamic, chemical, or spatial characterization is required

When both polar and non-polar metabolites are of interest, a biphasic extraction using methanol, chloroform, and water (2: 2:1.8, v/v/v) may be performed. After thorough mixing and centrifugation, two layers form. The upper aqueous layer, which contains polar metabolites, is carefully collected and dried completely before analysis.

3.4.5 Sample Preparation for NMR

The dried metabolite extract is reconstituted in 600 μL of phosphate buffer prepared in deuterium oxide (D_2O). An internal reference compound, such as DSS (0.5 mM), is added to allow chemical shift calibration and quantitative analysis. The pH is adjusted to 7.0 ± 0.1 to minimize chemical shift variation between samples. The solution is transferred to a 5 mm NMR tube, ensuring no air bubbles remain, as these can interfere with spectral quality. Maintaining consistent sample volume and pH across all samples is critical for meaningful comparison.

3.4.6 NMR Data Acquisition

(a) Instrument Conditions

Spectra are acquired at 25 °C using standard one-dimensional ^{1}H NMR parameters. A relaxation delay of at least 5 s is applied to ensure adequate relaxation between scans. The spectral width is typically set to approximately 12 ppm, and 64–128 scans are collected to achieve a satisfactory signal-to-noise ratio. Water suppression techniques, such as presaturation or NOESY-presaturation, are used to reduce interference from residual water signals (*see* **Note 9**).

(b) Pulse Sequences

A standard 1D ^{1}H experiment with water suppression is used for routine metabolomic profiling. When additional structural confirmation is required, two-dimensional experiments such as COSY, HSQC, or TOCSY can be performed to clarify peak assignments and resolve overlapping signals.

3.4.7 Data Processing

Acquired free induction decay (FID) data are Fourier-transformed to generate spectra. Phase and baseline corrections are applied carefully to improve spectral accuracy. Chemical shifts are referenced to the internal standard (DSS at 0.00 ppm).

The residual water region (approximately 4.5–5.0 ppm) is excluded from analysis. Spectra are then normalized either to the total spectral area or to the internal standard signal to allow comparison between samples.

3.4.8 Metabolite Identification

Metabolites are assigned based on their chemical shift values, signal multiplicity, and coupling patterns. Reference databases such as the Human Metabolome Database (HMDB) and the Biological Magnetic Resonance Data Bank (BMRB) are used to confirm peak identities.

Compounds commonly observed in biofilm samples include organic acids (e.g., lactate and acetate), amino acids (such as alanine), carbohydrate derivatives, and sugars associated with extracellular polymeric substances (*see* **Notes 10** and **11**).

3.4.9 Multivariate Data Analysis

To compare metabolic differences between biofilm stages, spectra are segmented into small regions (binning, typically 0.01–0.04 ppm).

The resulting data matrix is normalized and subjected to statistical analysis.

Principal component analysis (PCA) is used to explore clustering patterns, while partial least squares discriminant analysis (PLS-DA) can be applied to distinguish specific groups, such as early-stage biofilm, mature biofilm, and planktonic controls.

3.4.10 Interpretation of Results

As biofilms mature, characteristic metabolic trends are often observed. Signals corresponding to extracellular carbohydrates typically increase, reflecting matrix production. Amino acid profiles may shift, and fermentation-related metabolites can accumulate over time. Concurrently, nutrient-related signals may decrease as substrates are consumed. These metabolic changes provide insight into physiological adaptation during biofilm development.

3.4.11 Quality Control Measures

To ensure data reliability, solvent blanks should be analyzed to detect contamination. A pooled quality control sample prepared from all extracts can help monitor instrument stability. Consistent shimming and temperature control are essential throughout data acquisition.

3.4.12 Statistical Analysis

All experiments should include at least three biological replicates. Statistical comparisons between groups may be performed using t-tests or one-way ANOVA, with significance typically set at $p < 0.05$. Multivariate models such as PLS-DA should be validated using cross-validation methods to avoid overfitting.

3.4.13 Expected Outcome

^{1}H NMR–based metabolomics provides a broad metabolic profile of biofilm samples, allowing quantitative assessment of metabolites and detailed insight into extracellular matrix composition.

The method offers a reproducible and non-destructive means of investigating biochemical changes during biofilm formation and maturation.

3.5 Biofilm Detection by Magnetic Resonance Imaging (MRI)

Magnetic resonance imaging (MRI) offers a noninvasive way to visualize biofilms in hydrated systems. Unlike microscopy, MRI does not require staining or destructive sampling and can monitor biofilm development inside opaque materials, reactors, or medical devices. The technique relies on differences in proton density and relaxation behavior between free water and biofilm-associated water, allowing structural mapping and growth tracking over time. MRI detects signals from hydrogen nuclei (^{1}H) in water molecules. In biofilms, water mobility is restricted by the extracellular polymeric substance (EPS) matrix and cellular biomass [7]. These changes influence signal intensity, T_1 and T_2 relaxation behavior, and diffusion properties. As a result, biofilm regions can be distinguished from surrounding bulk fluid in MR images. MRI is particularly useful for:

- Monitoring biofilm growth over time
- Measuring thickness and spatial distribution
- Observing structural heterogeneity
- Studying biofilms in flow systems or porous materials

3.5.1 Instrumentation and Materials

- High-field or low-field MRI-compatible magnetic resonance imaging scanner
- Small-bore imaging system (for laboratory-scale studies)
- Radiofrequency (RF) coil suitable for sample size
- Temperature control unit
- Biofilm-forming microorganism
- Sterile growth medium
- Flow cell, tubing, or imaging chamber compatible with MRI
- Phosphate-buffered saline (PBS)

3.5.2 Biofilm Growth Preparation

1. Inoculate the selected biofilm-forming bacterial strain into sterile growth medium.
2. Incubate overnight at optimal temperature (commonly 37 °C) with agitation.
3. Transfer diluted culture into a sterile flow cell, glass tube, or porous material that fits inside the MRI scanner.
4. Allow biofilm formation under static or controlled flow conditions.
5. Maintain sterile conditions to avoid contamination.

For longitudinal studies, biofilm growth can be monitored at intervals such as 24, 48, and 72 h without disturbing the structure.

3.5.3 Sample Positioning and Setup

1. Carefully place the biofilm-containing chamber into the MRI coil.
2. Ensure proper alignment within the magnetic field.
3. Remove air bubbles, as they cause susceptibility artifacts in MR images.
4. Stabilize temperature before scanning to maintain consistent relaxation behavior.

3.5.4 MRI Acquisition Parameters

(a) Imaging Sequence

A T_2-weighted spin-echo sequence is commonly used for biofilm detection because it highlights differences in water mobility between biofilm and bulk fluid (*see* **Notes 12** and **13**).

Typical parameters:

- Repetition time (TR): 1500–3000 ms
- Echo time (TE): 20–80 ms
- Slice thickness: 0.5–2 mm
- Field of view (FOV): Adjusted to sample size
- Matrix size: 128 × 128 or higher for improved resolution

(b) Alternative Sequences

- T_1-weighted imaging for contrast variation
- Diffusion-weighted imaging (DWI) to assess restricted diffusion
- 3D imaging for volumetric biofilm reconstruction

3.5.5 Image Acquisition

1. Perform scout scan to localize sample.
2. Acquire baseline image before biofilm formation (control).
3. Collect sequential images at defined growth intervals.
4. Save raw image data for post-processing.

Repeated imaging allows monitoring of biofilm thickness and spatial expansion.

3.5.6 Image Processing and Analysis

1. Import images into image analysis software (e.g., ImageJ or manufacturer software).
2. Adjust contrast and normalize signal intensity.
3. Segment biofilm regions based on signal differences.
4. Measure biofilm thickness, surface coverage, and volume.
5. Compare signal intensity between biofilm and bulk medium.

Quantitative metrics may include:

- Biofilm thickness (mm)
- Total biofilm volume
- Signal intensity ratio (biofilm vs. water)

3.5.7 Validation of Biofilm Detection

To confirm MRI findings:

- Perform biomass staining (e.g., crystal violet assay)
- Use microscopy for structural confirmation
- Compare MRI thickness measurements with physical measurements

3.5.8 Quality Control

- Perform system calibration before imaging
- Maintain constant temperature
- Avoid magnetic interference
- Ensure identical imaging parameters across time points
- Include sterile medium control sample

3.5.9 Interpretation of Results

Biofilm regions typically appear with altered signal intensity relative to surrounding fluid due to restricted water mobility and matrix composition. As biofilm matures:

- Thickness increases
- Structural heterogeneity becomes visible
- Signal contrast between biofilm and fluid becomes more pronounced

MRI allows visualization of internal architecture without disturbing the sample, making it especially valuable for studying biofilms in porous materials, medical tubing, or bioreactors (*see* **Note 14**).

3.5.10 Expected Outcome

- MRI-based detection provides:
  - Noninvasive structural imaging
  - Real-time monitoring capability
  - Quantitative measurement of biofilm growth
  - Visualization inside opaque systems (*see* **Notes 15** and **16**).

4 Notes

1. When no significant change in T_2 is observed over time, the issue may be attributed to a weak or poorly developed biofilm, inconsistent growth conditions between samples, or the presence of excess bulk liquid that masks the biofilm signal. To address this, the incubation period should be extended to allow sufficient biofilm maturation. Biofilm formation should also be verified using microscopy or appropriate staining techniques. Additionally, excess PBS should be carefully reduced while ensuring the biofilm remains hydrated, and identical culture conditions must be maintained across all replicates to ensure experimental consistency and reliable T_2 measurements.
2. When the NMR relaxometry data show a weak or noisy signal, several technical factors may be responsible. In many cases, inadequate shimming of the magnetic field can reduce signal quality, while using too little sample volume may limit signal

intensity. The presence of air bubbles within the sample can further interfere with accurate measurements by disrupting magnetic field uniformity. To improve data quality, the magnetic field should be carefully re-shimmed to achieve optimal homogeneity, and a consistent sample volume should be used for all measurements. Samples should also be inspected to ensure that no air bubbles are present before placement in the instrument. If necessary, increasing the number of scans can help enhance the signal-to-noise ratio and produce more reliable results.

3. Variability in T_1 and T_2 values between replicates is often linked to inconsistencies in sample handling and environmental conditions. Differences in preparation procedures, fluctuations in temperature during analysis, or non-uniform biofilm thickness can all contribute to inconsistent relaxation times. Since relaxation parameters are sensitive to temperature, maintaining a stable and controlled environment throughout the experiment is essential. It is equally important to standardize washing and preparation protocols to reduce procedural variation. Performing experiments in biological triplicates can help account for natural variability, and special care should be taken during sample transfer to avoid disturbing the biofilm structure, as any disruption may directly affect the measured relaxation values.
4. When the calculated diffusion coefficient values appear inconsistent or physically unrealistic, the issue is often linked to instrumental or experimental factors. One common cause is inaccurate gradient calibration, which can directly affect the accuracy of diffusion measurements. Convection within the sample may also introduce artifacts, particularly in liquid systems, leading to misleading diffusion values. In some cases, sample overheating during acquisition can further distort the results. To address these problems, the gradient system should be recalibrated to ensure accurate gradient strength application.
5. Employing a convection-compensated pulse sequence can help minimize flow-related artifacts, and the sample temperature should be allowed to stabilize fully before starting data acquisition.
6. If signal attenuation during diffusion experiments is either too weak or excessively strong, the gradient parameters may not be properly optimized. An unsuitable gradient strength range can result in insufficient or overly severe signal decay, making accurate fitting difficult. Similarly, inappropriate choices of gradient pulse duration (δ) or diffusion time (Δ) can compromise the measurement. To improve the quality of attenuation data, the

gradient increment range should be carefully adjusted to achieve a suitable spread of signal decay. Diffusion delay parameters should be optimized based on the expected diffusion coefficient, and the repetition time (TR) must be sufficiently long to allow complete relaxation between scans.

7. Unstable or unreliable multi-exponential fitting is frequently associated with low signal-to-noise ratios or the use of overly complex models. When data quality is poor, fitting multiple components can lead to non-physical or highly variable results.
8. In addition, applying an overly complex model may cause overfitting, reducing the reliability of the extracted parameters. Increasing the number of scans can improve signal quality and stabilize the fitting process. It is often advisable to begin with a simpler mono-exponential model to establish baseline behavior before introducing more complex models. Finally, validating the fitting results through residual analysis can help confirm whether the chosen model adequately represents the experimental data.
9. When poor spectral resolution is observed in NMR analysis, the problem is often related to suboptimal magnetic field homogeneity, unsuitable sample conditions, or contamination. Inadequate shimming can broaden peaks and reduce spectral clarity, while incorrect pH may alter chemical shift positions and peak shapes. The presence of salts or other impurities can further compromise resolution by increasing line broadening. To improve spectral quality, the magnet should be carefully reshimmed to achieve optimal field uniformity. The sample pH should be adjusted to 7.0 ± 0.1 to ensure consistency, and extracts should be filtered or purified if necessary. Using the same buffer system for all samples also helps maintain reproducibility and spectral stability.
10. A dominant water peak can sometimes obscure nearby metabolite signals, making spectral interpretation difficult. This typically results from inefficient water suppression or incorrect presaturation settings. To minimize water interference, suppression parameters should be carefully optimized for the specific sample type. Implementing a NOESY-presat pulse sequence can improve water suppression efficiency, and adjusting the presaturation power level and frequency offset can further reduce residual water signals without affecting metabolite peaks.
11. If noticeable peak shifts occur between samples, variations in experimental conditions are usually responsible. Even small differences in pH can lead to measurable chemical shift

changes. Temperature fluctuations during acquisition can also influence resonance positions, and inconsistent concentrations of the internal reference compound may result in inaccurate calibration. To avoid these issues, all samples should be adjusted to the same pH, measured under controlled and constant temperature conditions, and prepared with identical concentrations of the internal standard.

12. Poor metabolite extraction efficiency is often linked to inadequate sample handling during the extraction process. Incomplete metabolic quenching may allow ongoing biochemical activity, altering metabolite composition. Insufficient mixing can reduce extraction yield, and prolonged processing times may cause sample degradation. To enhance extraction efficiency, cold methanol should be added immediately after sampling to rapidly quench metabolism. Thorough vortexing or mixing is essential to ensure effective metabolite recovery, and samples should be kept cold throughout the procedure. Reducing the total processing time further helps preserve metabolite integrity and improve reproducibility.
13. Image artifacts, such as geometric distortion or localized signal voids, are commonly encountered in MRI-based biofilm studies. These issues are frequently caused by trapped air bubbles, differences in magnetic susceptibility between materials, or improper sample positioning within the radiofrequency coil. Air bubbles can create strong local field inhomogeneities, leading to signal loss or distortion. Similarly, susceptibility differences between the sample, container, and surrounding medium may disrupt magnetic field uniformity. To minimize these artifacts, any air bubbles should be carefully removed prior to scanning, and degassed media can be used to reduce bubble formation. Ensuring that the sample is correctly centered and properly positioned within the coil also helps maintain uniform signal acquisition.
14. Poor contrast between the biofilm and the surrounding bulk fluid can make structural visualization challenging. This often results from suboptimal imaging parameters, insufficient biofilm thickness, or inadequate echo time (TE) for proper T_2 weighting. If TE is too short, differences in transverse relaxation between the biofilm and fluid may not be sufficiently emphasized. To enhance contrast, the echo time can be increased to strengthen T_2 weighting, and a dedicated T_2-weighted imaging sequence should be selected. In cases where the biofilm is too thin to produce measurable contrast, extending the growth period may improve detectability.

15. Motion artifacts are particularly problematic in flow-based systems, where fluid movement during scanning can cause blurring or ghosting in the images. External vibrations may further degrade image quality. To reduce motion-related distortions, fluid flow should be paused during image acquisition whenever possible. Tubing and other components should be securely fixed to prevent mechanical movement. Additionally, motion-compensated imaging sequences can be employed to minimize the impact of residual motion.

16. Low spatial resolution may limit the ability to visualize fine biofilm structures. This is often due to the use of thick imaging slices or a small acquisition matrix. Improving spatial resolution can be achieved by reducing slice thickness and increasing the matrix size, thereby enhancing image detail. If available, three-dimensional imaging techniques can provide higher-resolution datasets and more accurate structural representation of the biofilm.

Temperature Instability

- Always allow system equilibration before measurement (Table 2)
- Use temperature control unit consistently

Sample Variability

- Use biological triplicates
- Standardize incubation time and conditions

Reproducibility Issues

- Maintain consistent buffer composition
- Use identical NMR parameters for all runs
- Document every parameter carefully

Magnetic Interference

- Keep ferromagnetic objects away
- Use MRI/NMR-compatible materials only

Table 2
Overall and common trouble shooting issues of all types of NMR

Method	Most common issue	Primary fix
Relaxometry	No T_2 change	Confirm biofilm formation
PFG-NMR	Incorrect diffusion values	Recalibrate gradients
^{1}H NMR metabolomics	Poor resolution	Adjust pH and shimming
MRI	Imaging artifacts	Remove air bubbles

References

1. Abreu AC, Fernández I (2020) Chapter 2 - Nuclear magnetic resonance to study bacterial biofilms structure, formation, and resilience, recent trends in biofilm science and technology. Academic Press, pp 23–70. https://doi.org/10.1016/B978-0-12-819497-3.00002-7
2. Zhang Y, Xu A, Lv X, Wang Q, Feng C, Lin J (2021) Non-invasive measurement, mathematical simulation and in situ detection of biofilm evolution in porous media: a review. Appl Sci 11(4):1391. https://doi.org/10.3390/app11041391
3. Byeon CH, Kinney T, Saricayir H, Holst Hansen K, Scott FJ, Srinivasa S, Wells MK, Mentink-Vigier F, Kim W, Akbey Ü (2025) Ultrasensitive characterization of native bacterial biofilms via dynamic nuclear polarization-enhanced solid-state NMR. Angew Chem Int Ed Engl 64(12):e202418146. https://doi.org/10.1002/anie.202418146
4. Srithabut S, Chareonsudjai P, Chaianunporn T, Bunma C, Phetcharaburanin J, Suksawat M, Chitcharoen S, Chareonsudjai S (2025) NMR-based metabolomics of *Burkholderia pseudomallei* biofilms and extracellular polymeric substance cultured in LB and MVBM media. Biofouling 41(5):498–511. https://doi.org/10.1080/08927014.2025.2502936
5. Leanne C, Garnett James A (2023) How to study biofilms: technological advancements in clinical biofilm research. Front Cell Infect Microbiol 13:1335389. https://doi.org/10.3389/fcimb.2023.1335389
6. Peng Y, Zhang Z, He L, Li C, Liu M (2024) NMR spectroscopy for metabolomics in the living system: recent progress and future challenges. Anal Bioanal Chem 416(9):2319–2334. https://doi.org/10.1007/s00216-024-05137-8
7. Willett MR, Codd SL, Seymour JD, Kirkland CM (2024) Relaxation-weighted MRI analysis of biofilm EPS: differentiating biopolymers, cells, and water. Biofilm 8:100235. https://doi.org/10.1016/j.bioflm.2024.100235

Chapter 20

Biofilm Detection and Profiling Using MALDI-TOF and Liquid Chromatography–Mass Spectrometry

Abirami K

Abstract

Biofilms represent complex microbial communities embedded within a self-produced extracellular matrix, contributing significantly to persistent infections, antimicrobial resistance, and industrial biofouling. Accurate detection and detailed characterization of biofilms are therefore critical for understanding their formation, composition, and pathogenic potential. Mass spectrometry has emerged as a powerful analytical approach for biofilm investigation due to its sensitivity, specificity, and ability to provide molecular-level insights. This chapter focuses on the application of matrix-assisted laser desorption/ionization time-of-flight mass spectrometry (MALDI-TOF MS) and liquid chromatography–mass spectrometry (LC–MS) in the detection and profiling of microbial biofilms. MALDI-TOF MS enables rapid identification of biofilm-associated microorganisms and differentiation between planktonic and biofilm states based on distinct protein fingerprinting patterns. In contrast, LC–MS offers comprehensive proteomic and metabolomic profiling, allowing detailed analysis of biofilm-specific biomarkers, extracellular polymeric substances, and metabolic adaptations. Together, these complementary techniques provide valuable insights into biofilm architecture, microbial interactions, and resistance mechanisms. The integration of MALDI-TOF and LC–MS holds significant promise for advancing biofilm diagnostics, monitoring biofilm development, and supporting targeted therapeutic and preventive strategies in clinical and industrial settings.

Key words Biofilm monitoring, Microbial community profiling, Environmental biofilms, Industrial biofouling, Proteomic profiling, Metabolomic analysis, Microbial identification

1 Introduction

Microbial biofilms represent structured communities of microorganisms embedded within a self-produced extracellular matrix. Cells in the biofilm state exhibit physiological and molecular features that differ significantly from planktonic cells. Mass spectrometry-based approaches are employed in this study to detect and characterize biofilms at the molecular level. MALDI-TOF MS is used for rapid profiling and differentiation of biofilm-associated cells, while LC–MS was applied for in-depth analysis of biofilm-specific proteins and metabolites. This methodology describes the

Abirami K (ed.), *Microbial Biofilm Research: Comprehensive Protocols and Best Practices*, Springer Protocols Handbooks, https://doi.org/10.1007/978-1-0716-5535-1_20,

use of matrix-assisted laser desorption/ionization time-of-flight mass spectrometry (MALDI-TOF MS) and liquid chromatography–mass spectrometry (LC–MS) for the detection, identification, and molecular characterization of microbial biofilms. The techniques allow discrimination between planktonic and biofilm-associated cells and enable profiling of biofilm-specific proteins, peptides, and metabolites [1].

2 Materials

(a) Biological Materials.

- Clinical or laboratory microbial isolates (biofilm-forming organisms).
- Explanted medical devices (e.g., catheters, orthopedic implants, stents) *[for implant-associated studies].*
- Reference planktonic cultures (for comparative analysis).

(b) Culture Media and Buffers.

- Tryptic Soy Broth (TSB) / Brain Heart Infusion (BHI) / Luria–Bertani (LB) broth.
- Tryptic Soy Agar (TSA) or equivalent solid media.
- Phosphate-buffered saline (PBS), pH 7.2–7.4.
- Sterile distilled water (MS grade preferred).

(c) Biofilm Formation and Harvesting Materials.

- Sterile 96-well polystyrene microtiter plates.
- Glass coverslips or stainless-steel coupons.
- Sterile cell scrapers.
- Ultrasonic bath or probe sonicator (low-intensity setting).
- Sterile centrifuge tubes (1.5 mL, 2 mL, 15 mL).
- Wide-bore pipette tips.

(d) Reagents for MALDI-TOF MS Analysis.
Protein Extraction Reagents

- Absolute ethanol (≥99.5%).
- Formic acid (70%, MS grade).
- Acetonitrile (HPLC/MS grade).

(e) MALDI Matrix and Consumables.

- α-Cyano-4-hydroxycinnamic acid (HCCA).
- MALDI matrix solvent (acetonitrile/water with trifluoroacetic acid).
- Polished stainless-steel MALDI target plates.

(f) Reagents for LCMS Proteomic Analysis.
- Urea (ultrapure, proteomics grade).
- Protease inhibitor cocktail.
- Sequencing-grade trypsin.
- Dithiothreitol (DTT) *(optional for reduction).*
- Iodoacetamide (IAA) *(optional for alkylation).*
- Ammonium bicarbonate buffer.

(g) Reagents for LC–MS Metabolomic Analysis.
- Methanol (HPLC/MS grade).
- Water (LC–MS grade).
- Acetonitrile (LC–MS grade).
- Formic acid (≥99%).
- Sample cleanup and preparation materials.
- C18 solid-phase extraction (SPE) cartridges or spin columns.
- Vacuum centrifuge or nitrogen evaporator.
- 0.22 μm syringe filters (LC–MS compatible).
- LC–MS autosampler vials with inserts.

(h) Instrumentation.

MALDI-TOF MS
- MALDI-TOF mass spectrometer (e.g., Bruker Micro flex).
- Bacterial test standard (for calibration).

LC–MS
- UHPLC system.
- High-resolution mass spectrometer (Q-TOF or Orbitrap).
- Reverse-phase C18 analytical column.
- Electrospray ionization (ESI) source.

General Laboratory Equipment
- Refrigerated centrifuge (capable of ≥14,000× *g*).
- Microcentrifuge.
- Vortex mixer.
- Analytical balance.
- Ice bath.
- Biosafety cabinet (Class II).
- −20 °C and −80 °C freezers.
- Software and Data Analysis Tools.

3 Methods

3.1 Microorganism Culture and Biofilm Development for Mass Spectrometry

1. Microbial strains (e.g., *Staphylococcus aureus*, *Pseudomonas aeruginosa*, *E. coli*) are cultured overnight in suitable broth (e.g., TSB, LB, or BHI).
2. Cultures are adjusted to 0.5 McFarland standard (~1×10^8 CFU/mL).
3. Aliquots (200 μL) are inoculated into 96-well plates, glass slides, or stainless-steel coupons [2].
4. Incubate at 37 °C for 24–72 h under static conditions.
5. Media may be replaced every 24 h to enhance biofilm maturation (*see* **Note 1**).
6. Planktonic cells are removed by washing gently with phosphate-buffered saline (PBS).

3.1.1 Biofilm Harvesting

1. Biofilms are scraped using sterile cell scrapers or detached by sonication (low intensity, 30–60 s) in PBS.
2. Suspensions are centrifuged at 10,000× *g* for 10 min.
3. Pellets are washed twice with PBS.
4. Accurate mass spectrometric detection of biofilm-associated biomolecules requires careful harvesting of the biofilm biomass while preserving cellular integrity and minimizing contamination from extracellular media components.
5. The following protocol describes a standardized method for collecting and preparing biofilm samples for downstream mass spectrometry analysis.

3.1.2 Step 1: Removal of Planktonic Cells

1. The planktonic cells are removed after completion of the biofilm incubation period, and the culture medium containing planktonic cells is gently aspirated without disturbing the attached biofilm.
2. The biofilm-coated surface is then rinsed slowly with sterile phosphate-buffered saline (PBS, pH 7.2–7.4) to remove loosely attached and non-adherent cells. This washing step is repeated once to ensure that only firmly attached biofilm cells remain.

3.1.3 Step 2: Mechanical Detachment of Biofilm

Biofilms are detached from the surface using one of the following approaches:

1. Scraping method: A sterile cell scraper is used to gently scrape the biofilm from the surface into a small volume of PBS (*see* **Note 2**).

2. Sonication method: Alternatively, biofilm-coated substrates are submerged in PBS and subjected to low-intensity sonication for 30–60 s.
3. Sonication is performed under controlled conditions to disrupt the biofilm matrix while avoiding extensive cell lysis [3].

3.1.4 Step 3: Collection of Biofilm Suspension

1. The detached biofilm material is transferred into sterile microcentrifuge tubes using pipettes with wide-bore tips to prevent mechanical damage to cells.
2. The suspension is gently mixed to ensure homogeneity (*see* **Note 3**).

3.1.5 Step 4: Centrifugation

1. The biofilm suspension is centrifuged at 10,000× *g* for 10 min at 4 °C. This step pellets the biofilm cells and associated matrix components at the bottom of the tube.

3.1.6 Step 5: Removal of Supernatant

1. The supernatant is carefully discarded without disturbing the pellet after centrifugation.
2. Any residual liquid is removed using a fine pipette tip.

3.1.7 Step 6: Washing of Biofilm Pellet

1. The biofilm pellet is resuspended gently in sterile PBS by pipetting or brief vortexing. The suspension is centrifuged again at 10,000× *g* for 10 min.
2. This washing step is repeated twice.

3.1.8 Step 7: Final Pellet Preparation

1. After the final wash, the supernatant is discarded, and the cleaned biofilm pellet is retained.
2. The pellet can be
   (i) Immediately processed for protein or metabolite extraction, or
   (ii) Stored at −80 °C for short-term preservation prior to mass spectrometry analysis.

3.2 MALDI-TOF Mass Spectrometry (MALDI-TOF MS) Methodology

3.2.1 Protein Extraction from Biofilm Cells: Ethanol Formic Acid Extraction Method

1. Efficient extraction of intracellular proteins from biofilm-associated cells is essential for generating reproducible and high-quality MALDI-TOF MS spectra [4–7].
2. The ethanol formic acid extraction method was selected as it enhances protein solubilization while reducing interference from extracellular polymeric substances commonly present in biofilms (*see* **Note 4**).
3. The washed biofilm pellet obtained after harvesting is resuspended in 300 μL of sterile distilled water.
4. The suspension is mixed gently to ensure uniform distribution of cells.

5. To the resuspended cells, 900 μL of absolute ethanol is added.
6. The mixture is vortexed vigorously for 1 min to ensure complete mixing.
7. The ethanol-treated suspension is centrifuged at 13,000× *g* for 2 min at room temperature.
8. This results in the formation of a compact pellet containing precipitated proteins and cellular debris.
9. After centrifugation, the supernatant is carefully discarded. The pellet is allowed to air-dry at room temperature for 5–10 min to remove residual ethanol.
10. The dried pellet is resuspended in 50 μL of 70% formic acid and mixed thoroughly.
11. An equal volume (50 μL) of acetonitrile is added to the formic acid suspension.
12. The mixture is vortexed and centrifuged again at 13,000× *g* for 2 min.
13. Clear supernatant containing extracted proteins is carefully transferred to a fresh microcentrifuge tube and used immediately for MALDI-TOF MS analysis or stored at −20 °C for short-term use (*see* **Note 5**).

3.2.2 Sample Spotting on MALDI Target Plate

1. A 1 μL aliquot of the extracted protein solution is applied onto a clean, polished stainless-steel MALDI target plate.
2. The applied sample is allowed to air-dry at room temperature, forming a thin protein film on the target surface.
3. Once dry, the sample spot is overlaid with 1 μL of α-cyano-4-hydroxycinnamic acid (HCCA) matrix solution prepared in organic solvent.
4. The matrix-analyte mixture is allowed to crystallize at room temperature, forming homogeneous crystals.

3.2.3 MALDI-TOF MS Instrumental Analysis

1. The MALDI target plate is inserted into the mass spectrometer. The instrument is operated in positive ion linear mode, which is optimal for analyzing intact bacterial proteins.
2. External calibration is performed using a bacterial test standard to ensure mass accuracy across the selected mass range.
3. Mass spectra are acquired over a mass-to-charge (m/z) range of 2,000–20,000, which encompasses the majority of ribosomal and housekeeping proteins used for microbial profiling. For each sample, spectra are recorded in triplicate from different regions of the sample spot.

3.2.4 Spectral Analysis and Data Interpretation

1. Raw spectra are subjected to baseline subtraction, smoothing, and normalization to remove noise and improve peak resolution.
2. Spectra obtained from biofilm-derived samples are compared with spectra from planktonic cells grown under identical conditions.
3. Peak intensity data are analyzed using:
4. Principal component analysis (PCA) to visualize clustering and variance
5. Hierarchical cluster analysis to assess similarity between samples.
6. Spectra are matched against commercial or in-house reference databases. Biofilm-specific spectral features may be incorporated into custom libraries to improve identification accuracy.

3.3 MALDI-TOF MS Methodology for Clinical Isolates and Implant-Associated Biofilms

3.3.1 Collection and Pre-processing of Clinical Samples

Implant-associated biofilms are obtained from explanted medical devices such as intravascular catheters, orthopedic implants, or urinary stents under aseptic conditions. Immediately after removal, implants were placed into sterile containers containing phosphate-buffered saline (PBS) and transported to the laboratory for processing. All samples were handled in a biosafety cabinet to minimize contamination and ensure operator safety (Fig. 1) [8].

3.3.2 Detachment of Biofilm from Implant Surfaces

1. Implants were gently rinsed with sterile PBS to remove loosely adherent blood components and host-derived debris without disturbing the biofilm structure.
2. Biofilm detachment was performed by immersing the implant in sterile PBS followed by controlled sonication (30–60 s at low frequency). In cases where sonication was not feasible, biofilms were carefully scraped using sterile instruments.
3. The resulting biofilm suspension was transferred into sterile centrifuge tubes and gently mixed to ensure homogeneity.
4. The biofilm suspension was centrifuged at 10,000× *g* for 10 min at 4 °C. The supernatant was discarded, and the pellet was resuspended in sterile PBS. This washing step was repeated twice.

3.3.3 Protein Extraction Using Ethanol–Formic Acid Method

1. The washed biofilm pellet was resuspended in 300 μL of sterile distilled water and mixed gently.
2. To the suspension, 900 μL of absolute ethanol was added, followed by vigorous vortexing.

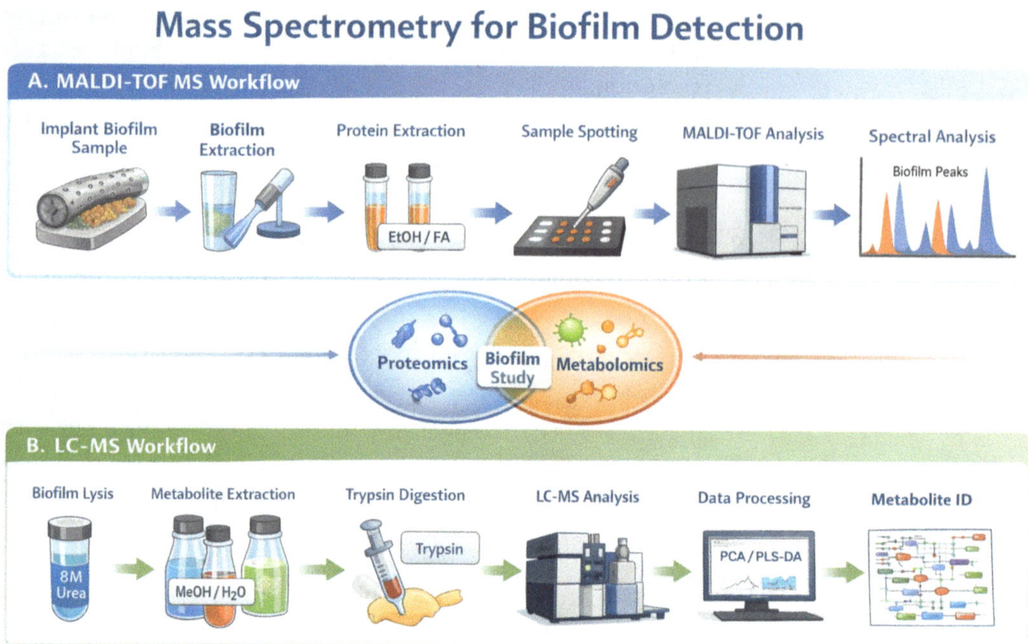

Fig. 1 (**a**) MALDI-TOF MS profiling of planktonic and biofilm cells. Mass spectra generated by matrix-assisted laser desorption/ionization time-of-flight mass spectrometry (MALDI-TOF MS) showing protein fingerprint differences between planktonic and biofilm-associated bacterial cells. Spectra are acquired in linear positive mode within the 2–20 kDa range. Distinct peak shifts and intensity variations indicate biofilm-specific protein expression. Representative spectra from three biological replicates are shown. (**b**) LC–MS-based proteomic characterization of biofilm matrix proteins. Proteomic analysis performed using liquid chromatography–mass spectrometry (LC–MS) to identify differentially expressed proteins in biofilm versus planktonic cells. Total proteins were extracted, enzymatically digested, and analyzed by LC–MS/MS. Identified proteins were annotated based on functional categories including adhesion, extracellular matrix production, stress response, and quorum sensing

3. The mixture was centrifuged at 13,000× *g* for 2 min. The supernatant was discarded, and the pellet was allowed to air-dry.
4. The dried pellet was resuspended in 50 μL of 70% formic acid and thoroughly mixed.
5. An equal volume (50 μL) of acetonitrile was added, followed by vortexing and centrifugation at 13,000× *g* for 2 min.

3.3.4 Sample Spotting and Matrix Application

1. A 1 μL aliquot of the extracted protein solution was spotted onto a clean MALDI target plate.
2. The sample spot was air-dried at room temperature to allow analyte concentration.
3. Each spot was overlaid with 1 μL of α-cyano-4-hydroxycinnamic acid (HCCA) matrix solution and allowed to crystallize.

3.4 MALDI-TOF MS Instrumental Conditions

- Instrument: Clinical MALDI-TOF MS system (e.g., Bruker Micro flex)
- Ion mode: Positive linear mode
- Mass range: 2,000–20,000 m/z
- Calibration: External calibration using bacterial test standard
- Acquisition: Spectra recorded in triplicate from different positions on each spot

3.4.1 Spectral Analysis and Interpretation

1. Raw spectra were subjected to baseline correction, smoothing, and normalization.
2. Spectra obtained from implant-associated biofilms were compared with:
   - Planktonic clinical isolates.
   - Reference spectra from commercial databases.
3. Spectral datasets were analyzed using:
   - Principal component analysis (PCA)
   - Hierarchical cluster analysis
4. These methods enabled discrimination between planktonic and biofilm-associated clinical isolates. Species identification was achieved by database matching. Biofilm-specific peaks were further analyzed to explore associations with antimicrobial resistance and chronic implant-related infections.

   Clinical Significance:

This adapted MALDI-TOF MS methodology enables rapid identification and differentiation of microorganisms involved in implant-associated biofilm infections, providing valuable insights for diagnosis, treatment selection, and infection control strategies.

3.5 Liquid Chromatography–Mass Spectrometry (LC–MS) Methodology

3.5.1 Extraction of Proteins and Metabolites from Biofilm Cells

1. Biofilm-associated microorganisms exhibit distinct proteomic and metabolomic profiles compared to planktonic cells. Separate extraction workflows were therefore employed for protein-based (proteomics) and small-molecule-based (metabolomics) LC–MS analyses [9, 10].
2. Biofilm cell pellets were resuspended in a lysis buffer containing 8 M urea supplemented with a protease inhibitor cocktail. The suspension was gently mixed to ensure complete exposure of cells to the lysis buffer.
3. The lysate was sonicated on ice using short pulses to mechanically disrupt remaining cell structures. Cooling intervals were maintained to prevent heat-induced protein denaturation.
4. The lysed samples were centrifuged at 14,000× *g* for 15 min at 4 °C. Insoluble debris was pelleted, and the clarified supernatant containing soluble proteins was carefully collected.

5. Protein concentration was determined using a Bradford or bicinchoninic acid (BCA) assay following the manufacturer's instructions.
6. Biofilm pellets were resuspended in a cold solvent mixture of methanol: water (80:20, v/v) and vortexed vigorously.
7. Samples were incubated at −20 °C for 30 min to enhance metabolite extraction and suppress enzymatic activity.
8. Extracted samples were centrifuged at 14,000× *g* for 10 min. The metabolite-rich supernatant was transferred to fresh tubes (*see* **Note 6**).

3.5.2 Sample Cleanup and Preparation for LC–MS

1. Protein samples were reduced, alkylated if required, and digested with sequencing-grade trypsin at 37 °C overnight.
2. Digested peptides were purified using C18 solid-phase extraction cartridges to remove salts, detergents, and urea.
3. Purified peptides were dried under a gentle stream of nitrogen or using a vacuum centrifuge.
4. Dried peptides were reconstituted in the initial LC mobile phase, typically water containing 0.1% formic acid.
5. Metabolite extracts were dried and reconstituted in an appropriate LC solvent depending on chromatographic conditions (*see* **Note 6**).

3.5.3 LC–MS Instrumentation and Analytical Conditions

1. Samples were analyzed using an ultra-high-performance liquid chromatography (UHPLC) system equipped with a C18 reverse-phase column. Separation was achieved using a binary solvent system:
2. Mobile Phase A: Water with 0.1% formic acid.
3. Mobile Phase B: Acetonitrile with 0.1% formic acid.
4. A linear gradient from 5% to 95% mobile phase B was applied over 30–60 min.
5. The UHPLC system was coupled to a high-resolution mass spectrometer (Q-TOF or Orbitrap) operated in electrospray ionization (ESI) mode.
6. Data-dependent MS/MS acquisition was employed to fragment selected precursor ions for structural identification (*see* **Note 7**).

3.5.4 Data Processing and Bioinformatic Analysis

1. Raw proteomic data were processed using MaxQuant or Proteome Discoverer, while metabolomic data were analyzed using XCMS or MetaboAnalyst (*see* **Note 8**).
2. Comparative analysis was conducted between biofilm and planktonic samples to identify differentially expressed proteins and metabolites.

3. Identified features were annotated using curated databases:
   - Proteins: UniProt
   - Metabolic pathways: KEGG
   - Metabolites: HMDB

3.6 Validation and Statistical Analysis

1. All experiments were conducted in biological triplicates to ensure reproducibility.
2. Quantitative data were analyzed using:
   - Student's t-test for two-group comparisons
   - ANOVA for multiple group comparisons
3. A p-value <0.05 was considered statistically significant.
4. Unsupervised and supervised statistical approaches were applied:
   - Principal component analysis (PCA) to visualize clustering
   - Partial least squares discriminant analysis (PLS-DA) to identify discriminatory features

This LC–MS workflow enables high-resolution characterization of biofilm-associated proteomic and metabolomic alterations, providing mechanistic insights into biofilm formation, persistence, and antimicrobial resistance.

4 Notes

1. Biofilm analysis using MALDI-TOF MS and LC–MS is often technically demanding because environmental and industrial biofilms are structurally complex and highly heterogeneous. Variations in surface coverage, biofilm thickness, and biomass can significantly affect reproducibility, particularly during sample collection.
2. Ensuring consistent sampling areas and quantifying biomass prior to extraction are critical steps to minimize variability and obtain reliable results.
3. Sample preparation presents additional challenges. The dense extracellular polymeric substance (EPS) matrix, along with salts and environmental contaminants, can interfere with protein and metabolite extraction and suppress ionization during mass spectrometric analysis. Incomplete cell lysis may further reduce analytical sensitivity.
4. Employing a combination of mechanical disruption and chemical lysis, followed by appropriate cleanup procedures such as centrifugation, filtration, or solid-phase extraction, can substantially improve recovery and data quality. Effective desalting is particularly important when working with environmental samples.

5. In MALDI-TOF MS analysis, poor-quality spectra or inconsistent protein fingerprints are frequently linked to improper matrix crystallization, insufficient sample concentration, or residual contaminants. Careful optimization of matrix preparation, routine instrument calibration, and analysis of replicate spots enhance spectral consistency and confidence in identification.
6. LC–MS workflows may encounter issues such as inadequate chromatographic separation, ion suppression, and low identification rates, especially when analyzing complex microbial communities.
7. Refining gradient conditions, selecting suitable chromatographic columns, optimizing enzymatic digestion, and utilizing comprehensive microbial or metaproteomic databases can markedly improve detection and profiling outcomes.
8. Instrument performance also plays a crucial role in data reliability. Regular calibration, preventive maintenance, and the inclusion of blank runs help prevent mass drift and carryover contamination.
9. Ultimately, given the complexity of biofilm-derived datasets, the use of robust statistical and bioinformatics approaches is essential for accurate interpretation and meaningful biological insights.

Competing Interest

The author(s) declare that they have no financial interests that could be perceived as influencing the content of this chapter.

References

1. Parmar D, Rosado-Rosa JM, Shrout JD, Sweedler JV (2024) Metabolic insights from mass spectrometry imaging of biofilms: a perspective from model microorganisms. Methods 224:21–34. https://doi.org/10.1016/j.ymeth.2024.01.014
2. Çağatay İT (2024) Use of proteomic-based MALDI-TOF mass spectra for identification of bacterial pathogens in aquaculture: a review. Aquac Int 32:7835–7871. https://doi.org/10.1007/s10499-024-01544-x
3. Yuan L, Dai H, He G, Yang Z, Jiao X (2023) Invited review: current perspectives for analyzing the dairy biofilms by integrated multiomics. J Dairy Sci 106(12):8181–8192. https://doi.org/10.3168/jds.2023-23306
4. Tejan N, Fatima N, Yaduvanshi N, Singh R, Pathak A, Hasan I, Patel SS, Sahu C (2025) Evaluation of direct microbial identification by MALDI-TOF MS and antimicrobial susceptibility testing for early diagnosis of blood stream infections. BMC Microbiol 25(1):724. https://doi.org/10.1186/s12866-025-04401-w
5. Liu Y, Wu SY, Deng J (2024) Application of MALDI-TOF mass spectrometry for identification of *Nocardia* species. BMC Microbiol 24:358. https://doi.org/10.1186/s12866-024-03483-2
6. Cruz S, Abreu D, Gomes R, Martins-Oliveira I, Silva-Dias A, Perez-Viso B, Cantón R, Pina-Vaz C (2024) An improved protocol for

bacteria identification by MALDI-TOF MS directly from positive blood cultures. Eur J Clin Microbiol Infect Dis 43(3):605–610. https://doi.org/10.1007/s10096-023-04725-3

7. Viboud G, Asaro H, Huang MB (2024) Use of matrix-assisted laser desorption ionization time of flight (MALDI-TOF) to detect antibiotic resistance in bacteria: a scoping review. Am J Clin Pathol 161(4):317–328. https://doi.org/10.1093/ajcp/aqad160

8. Elbehiry A, Aldubaib M, Abalkhail A, Marzouk E, ALbeloushi A, Moussa I, Ibrahem M, Albazie H, Alqarni A, Anagreyyah S, Alghamdi S, Rawway M (2022) How MALDI-TOF mass spectrometry technology contributes to microbial infection control in healthcare settings. Vaccines (Basel) 10(11):1881. https://doi.org/10.3390/vaccines10111881

9. Dai Y, Xu X, Yan X, Li D, Cao W, Tang L, Hu M, Jiang C (2021) Evaluation of a rapid and simplified protocol for direct identification of microorganisms from positive blood cultures by using matrix assisted laser desorption ionization time-of-flight mass spectrometry (MALDI-TOF MS). Front Cell Infect Microbiol 11:632679. https://doi.org/10.3389/fcimb.2021.632679

10. Gaudreau AM, Labrie J, Goetz C, Dufour S, Jacques M (2018) Evaluation of MALDI-TOF mass spectrometry for the identification of bacteria growing as biofilms. J Microbiol Methods 145:79–81. https://doi.org/10.1016/j.mimet.2018.01.003

FSC
www.fsc.org
MIX
Papier | Fördert gute Waldnutzung
FSC® C083411

Kolibri 360 GmbH
Ferdinand-Jühlke-Straße 7
99095 Erfurt, Deutschland
produktsicherheit@kolibri360.de